SpringerBriefs in Energy

SpringerBriefs in Energy presents concise summaries of cutting-edge research and practical applications in all aspects of Energy. Featuring compact volumes of 50 to 125 pages, the series covers a range of content from professional to academic. Typical topics might include:

- A snapshot of a hot or emerging topic
- A contextual literature review
- A timely report of state-of-the art analytical techniques
- An in-depth case study
- A presentation of core concepts that students must understand in order to make independent contributions.

Briefs allow authors to present their ideas and readers to absorb them with minimal time investment.

Briefs will be published as part of Springer's eBook collection, with millions of users worldwide. In addition, Briefs will be available for individual print and electronic purchase. Briefs are characterized by fast, global electronic dissemination, standard publishing contracts, easy-to-use manuscript preparation and formatting guidelines, and expedited production schedules. We aim for publication 8–12 weeks after acceptance.

Both solicited and unsolicited manuscripts are considered for publication in this series. Briefs can also arise from the scale up of a planned chapter. Instead of simply contributing to an edited volume, the author gets an authored book with the space necessary to provide more data, fundamentals and background on the subject, methodology, future outlook, etc.

SpringerBriefs in Energy contains a distinct subseries focusing on Energy Analysis and edited by Charles Hall, State University of New York. Books for this subseries will emphasize quantitative accounting of energy use and availability, including the potential and limitations of new technologies in terms of energy returned on energy invested. The second distinct subseries connected to SpringerBriefs in Energy, entitled Computational Modeling of Energy Systems, is edited by Thomas Nagel, and Haibing Shao, Helmholtz Centre for Environmental Research - UFZ, Leipzig, Germany. This sub-series publishes titles focusing on the role that computer-aided engineering (CAE) plays in advancing various engineering sectors, particularly in the context of transforming energy systems towards renewable sources, decentralized landscapes, and smart grids.

All Springer brief titles should undergo standard single-blind peer-review to ensure high scientific quality by at least two experts in the field.

Alfredo Iranzo · A. M. Kannan · Rafiq Ahmed ·
Christian Suárez · Felipe Rosa ·
Omkar Champhekar · Clemens Fink

Computational Fluid Dynamics Modelling of PEM Fuel Cells

From Theory to Practice

Springer

Alfredo Iranzo
ENGREEN—Laboratory of Engineering
for Energy and Environmental
Sustainability, Department of Energy
Engineering, Higher Technical School
of Engineering
University of Sevilla
Seville, Spain

Rafiq Ahmed
The Polytechnic School, Ira A. Fulton
Schools of Engineering
Arizona State University
Mesa, AZ, USA

Felipe Rosa
ENGREEN—Laboratory of Engineering
for Energy and Environmental
Sustainability, Department of Energy
Engineering, Higher Technical School
of Engineering
University of Sevilla
Seville, Spain

Clemens Fink
AVL List GmbH
Advanced Simulation Technologies
Graz, Austria

A. M. Kannan
The Polytechnic School, Ira A. Fulton
Schools of Engineering
Arizona State University
Mesa, AZ, USA

Christian Suárez
Department of Energy Engineering
University of Sevilla
Seville, Spain

Omkar Champhekar
Ansys, Inc.
Canonsburg, PA, USA

ISSN 2191-5520 ISSN 2191-5539 (electronic)
SpringerBriefs in Energy
ISBN 978-3-032-06630-5 ISBN 978-3-032-06631-2 (eBook)
https://doi.org/10.1007/978-3-032-06631-2

This Springer imprint is published by the registered company Springer Nature Switzerland AG
The registered company address is: Gewerbestrasse 11, 6330 Cham, Switzerland

If disposing of this product, please recycle the paper.

Preface

It is already over 30 years since the early journal publications and scientific reports appeared in the nineties on Fuel Cells (as an example, International Energy Agency, SOFC stack design tool, Final Rep., Swiss Federal Office of Energy, Berne, November 1992 [1]). During this time, a tremendous development has been achieved in many aspects including the fuel cell technology itself, the computing power, and the CFD physical models and solver algorithms. All this has led to today's achievements in fuel cell modelling capabilities ranging from the micro-modelling of gas diffusion in catalyst layers or gas diffusion layers, up to full stack modelling, and also including add-on features such as built-in degradation models.

The authors have been working on PEM Fuel Cell CFD during the last 15 years and have compiled in this book the main characteristics of the related CFD models, the unique features of the main CFD PEMFC commercial software, and the practical considerations and hints coming from their experience in real applications both in Academia and conducted for industry. Authors sincerely hope this book will become a useful tool for all researchers and practitioners involved in this thrilling and active field of work.

Seville, Spain
Mesa, USA
Mesa, USA
Seville, Spain
Seville, Spain
Canonsburg, USA
Graz, Austria

Alfredo Iranzo
A. M. Kannan
Rafiq Ahmed
Christian Suárez
Felipe Rosa
Omkar Champhekar
Clemens Fink

Acknowledgements Authors are sincerely thankful to all colleagues who somehow contributed to this book, in particular Manju Ramanathan, Project Coordinator at Springer Nature, for the kind support offered to fulfil any need regarding the book. Ms. Angeles Racero at the University of Seville kindly drafted many of the model equations in Chap. 4 authors are sincerely thankful for this valuable contribution. Authors are really thankful to all partners and colleagues, particularly to Ed Fontes and Henrik Ekström for their participation in Chap. 4, and sincerely hope that this book is fulfilling the expectations and provided a useful guide on this challenging topic.

Competing Interests The authors have no competing interests to declare that are relevant to the content of this manuscript.

Contents

Abbreviations/Nomenclature

Nomenclature

a	Water activity (–)
A	Area (m^2)
D	Diffusion coefficient (m^2/s)
E	Reversal potential (V)
EW	Equivalent weight (–)
F	Faraday constant (°C/kmol)
j	Exchange current density per active surface area (A/m^2)
K	Permeability (m^2)
M	Molecular weight (Kg/mol)
p	Pressure (Pa)
R	Current density Eq. 2–3 (A)
R	Universal gas constant Eq. 4–5 (Pa· m^3/mol·°C)
s	Liquid saturation (–)
S	Source (–)
t	Time (s)
T	Temperature (°C)
V	Volume (m^3)

Greek Symbols Description

α	Charge transfer coefficient (–)
γ	Concentration dependence coefficient (–)
δ	Thickness (m)
ε	Porosity (–)
η	Surface overpotential (–)
λ	Dissolved water content (–)
ρ	Density (kg/m^3)

Φ Transported variable ($-$)
Γ Diffusivity (m^2/s)
ζ Specific active surface area ($1/m$)
θ Phase-change rate exponent ($-$)
μ Dynamic viscosity ($Pa \cdot s$)

Abbreviation Definitions

CFD   Computational Fluid Dynamics
CL   Catalyst Layer
GDL   Gas Diffusion Layer
MEA   Membrane Electrode Assembly
PEM   Proton Exchange Membrane

Subscript Definitions

a   Anode
c   Cathode
cap   Capillary
eq   Equilibrium
g   Gas phase
gd   Gas dissolved
i   Porous media
l   Liquid phase
ld   Liquid dissolved
m   Membrane
osm   Osmotic
r   Relative
ref   Reference
w   Water
wv   Water vapor

Chapter 1
Introduction

Proton Exchange Membrane Fuel Cells (PEMFCs) represent a significant advancement in the field of electrochemical energy conversion. These fuel cells utilize hydrogen and oxygen as inputs and convert them into electrical energy, water, and heat through an electrochemical reaction. The most remarkable feature of PEMFCs is their clean operation—producing no greenhouse gas emissions, which is a crucial advantage in the quest for sustainable energy solutions. Unlike traditional internal combustion engines, PEMFCs offer a more environmentally friendly alternative with the potential to significantly reduce carbon footprints. PEMFCs operate at relatively low temperatures, typically up to about 90 °C, which contributes to their suitability for various applications, particularly in transportation. Their ability to generate electricity efficiently and with minimal emissions makes them particularly attractive for use in vehicles such as cars, buses, and even airplanes. Additionally, PEMFCs are highly scalable; they can be designed to meet a range of power outputs, from small, portable units to larger systems capable of powering significant vehicles or stationary applications.

The appeal of PEMFCs lies in their numerous advantages. One of the primary benefits is their operational efficiency. PEMFCs convert chemical energy directly into electrical energy through an electrochemical reaction, bypassing the intermediate step of combustion found in traditional engines. This direct conversion process results in higher efficiency and lower heat loss. Another significant advantage is the minimal environmental impact. PEMFCs produce only water and heat as by-products, making them a clean alternative to fossil fuel-based energy sources. The reduction in greenhouse gas emissions is a key factor in addressing climate change and improving air quality, particularly in urban environments where vehicle emissions contribute to pollution. Additionally, PEMFCs offer the potential for rapid refueling compared to battery electric vehicles. While battery electric vehicles can take several hours to charge, PEMFC vehicles can be refueled with hydrogen in just

A. Iranzo et al., *Computational Fluid Dynamics Modelling of PEM Fuel Cells*,
SpringerBriefs in Energy, https://doi.org/10.1007/978-3-032-06631-2_1

a few minutes. This rapid refueling capability is a crucial advantage in transportation applications, where downtime for refueling can significantly impact operational efficiency.

Despite their advantages, PEMFCs face several challenges that must be addressed before they can become a mainstream technology. One of the primary challenges is the cost. Currently, PEMFCs are expensive to produce, primarily due to the use of precious metals such as platinum in the catalysts. Reducing the cost of these materials or finding alternative catalysts is essential for making PEMFC technology more affordable and competitive with traditional internal combustion engines. Another challenge is the durability and longevity of PEMFCs. Fuel cells must withstand various operating conditions and remain reliable over extended periods. Issues such as degradation of the membrane, catalyst poisoning, and performance loss over time must be addressed to ensure the long-term viability of PEMFCs. Additionally, the infrastructure for hydrogen production, distribution, and storage is still underdeveloped compared to that for fossil fuels. The availability and accessibility of hydrogen refueling stations are limited, which poses a significant barrier to the widespread adoption of PEMFC vehicles. Developing a robust hydrogen infrastructure is crucial for the success of PEMFC technology in the transportation sector.

To overcome these challenges and enhance the performance of PEMFCs, researchers and engineers employ various tools and techniques, one of which is Computational Fluid Dynamics (CFD) modeling. CFD is a branch of fluid mechanics that uses numerical analysis and algorithms to solve and analyze problems involving fluid flows. In the context of PEMFCs, CFD modeling is used to simulate the behavior of fluids within the fuel cell and optimize its design and performance. Computer-Aided Drawing (CAD): CAD is used to create detailed and accurate representations of the fuel cell components. This digital modeling is essential for setting up the simulation environment and ensuring that the geometries used in the CFD analysis are precise.

Sophisticated mathematical models are employed to represent the physical and chemical processes occurring within the fuel cell. These models include equations governing fluid flow, heat transfer, electrochemical reactions, and other relevant phenomena. Once the mathematical models are established, numerical techniques are used to solve the equations and obtain results. This involves discretizing the computational domain into smaller elements and applying numerical methods to solve the equations governing fluid dynamics and electrochemical processes. Several commercial software packages are available for PEMFC CFD modeling, each offering various features and capabilities. Some of the well-known software tools include ANSYS Fluent, Star-CCM+, AVL FIRE M, COMSOL Multiphysics, and others.

ANSYS Fluent is a widely used CFD software that provides advanced capabilities for modeling fluid dynamics, heat transfer, and chemical reactions. It is known for its robust solver algorithms and user-friendly interface, making it a popular choice for fuel cell simulations. AVL FIRE M is a general all-purpose multi-physics CFD software, mainly used in the automotive industry for the development of all types of powertrains and their components, including PEM fuel cell stacks, which is including additional features such as a cell degradation model. COMSOL Multiphysics is another powerful tool that allows for the coupling of various physical phenomena in a single simulation. It offers flexibility in modeling complex systems and is often used for detailed analysis of fuel cell performance. OpenFOAM is an open-source CFD toolbox that provides a range of solvers and utilities for fluid flow simulations. It is highly customizable and allows researchers to develop and implement their own models and algorithms. AVL CRUISE™ and GT-SUITE are multi-domain simulation software tools that includes capabilities for modeling fuel cells and their integration with other systems. They are used for comprehensive system-level simulations, including fuel cell performance and interactions with other vehicle components. While these software tools simplify some aspects of the modeling process, achieving accurate and reliable results remains a significant challenge. The complexity of the physical processes involved, combined with the need for detailed input data and validation, makes CFD modeling a demanding task. Proton Exchange Membrane Fuel Cells (PEMFCs) offer a promising alternative to traditional internal combustion engines, with their high efficiency, low emissions, and scalability. However, several challenges must be addressed to make PEMFC technology competitive and widely adopted. These challenges include reducing costs, improving durability, and developing a robust hydrogen infrastructure.

Computational Fluid Dynamics (CFD) modeling plays a crucial role in addressing these challenges by providing insights into the behavior of fluids and electrochemical processes within the fuel cell. Through the use of advanced software tools and numerical techniques, researchers and engineers can optimize fuel cell designs, enhance performance, and contribute to the advancement of PEMFC technology. As research and development in this field continue, it is anticipated that PEMFCs will become increasingly viable and widespread, contributing to a more sustainable and environmentally friendly future. The ongoing efforts to improve fuel cell performance, reduce costs, and develop hydrogen infrastructure will be pivotal in realizing the full potential of PEMFC technology.

Chapter 2
PEM Fuel Cell Technology

Fuel cells are combustion-free electrical energy generators that convert chemical energy from hydrogen to electrical energy using electrochemical redox reactions. As shown in Figure 2.1, hydrogen is a common fuel that passes through an anode, while an oxidizer (oxygen/air) is supplied at the cathode. Both electrodes are separated by a charge-selective electrolyte membrane that allows only a specific charge species to pass through it. The byproducts are heat, water, and electricity. The fuel cell is divided into five subcategories according to the various parameters, such as the efficiency, power density and operating temperature, the details of which are listed in Table 2.1.

2.1 Alkaline Fuel Cell

As shown in Figure 2.1a, the alkaline fuel cell (AFC) is one of the first technologies developed and used in the U.S. Apollo 11 space program to generate electricity and water onboard spacecraft [1]. AFCs use a solution of potassium hydroxide in water as an electrolyte, and due to their mild corrosive nature, AFCs utilize a variety of nonprecious transition metals as catalysts at the cathode and anode [2]. These fuel cells use hydroxide (OH^-) exchange membranes as solid electrolytes to separate both electrodes and operate at 60–80 °C. Under these conditions, AFCs deliver an intermittent electrical efficiency of approximately 40%, while under optimum conditions (e.g., space applications), the electrical efficiency can reach up to 60% [3]. Due to CO_2 poisoning, minute traces of CO_2 in input gases can cause alkaline carbonates to solidify in the electrolyte and reduce the lifetime of commercial stacks; thus, the H_2 and O_2 gases supplied at the anode and cathode should be 99.9% pure. This purification and early replacement of cells due to CO_2 poisoning adds additional costs to AFC maintenance expenses. The overall cost of AFCs depends on the type of production, for example, in unit production for remote environments (space application or under sea operation); the cost is less of a matter. However, for commercial

A. Iranzo et al., *Computational Fluid Dynamics Modelling of PEM Fuel Cells*,
SpringerBriefs in Energy, https://doi.org/10.1007/978-3-032-06631-2_2

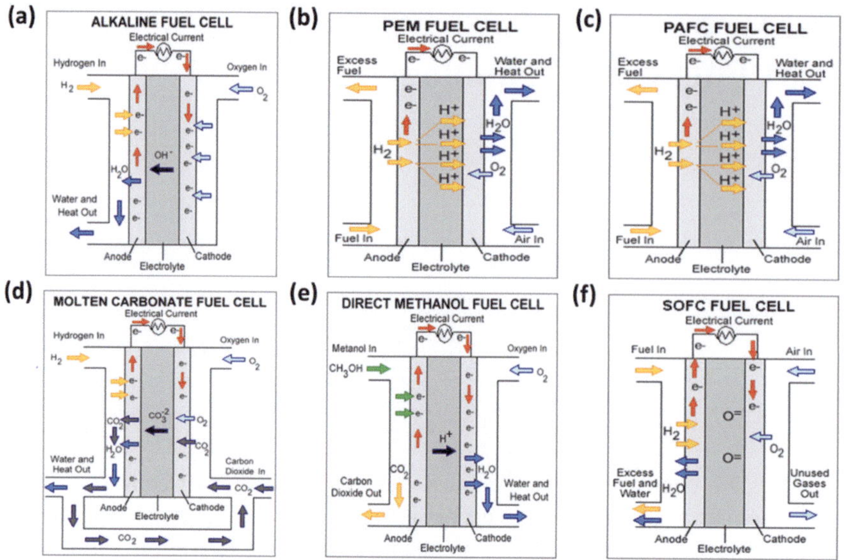

Fig. 2.1 Schematic diagram of fuel cells, **a**. AFC, **b**. PEMFC, **c**. PAFC, **d**. MCFC, **e**. DMFC and **f**. SOFCs

widespread market use, such as batch production and mass production, these fuel cells must undergo improvements to enhance cost effectiveness. In addition, the most significant problem that hurdles the commercialization of this technology is the low number of operation hours. For example, commercial AFCs have been tested for stable operation efficiency for 8,000 h [4]. However, this approach is insufficient for achieving economic viability in extensive utility applications. To compete with fossil fuel-based power technologies, they must surpass operating durations of 40,000 hours [5].

2.2 Proton Exchange Membrane Fuel Cell

Proton exchange membrane fuel cells, also known as polymer electrolyte membrane fuel cells (PEMFCs), are relatively new compared to AFCs and are still in the research development phase. As shown in Figure 2.1b, PEMFCs use a proton-conducting perfluorosulfonic acid electrolyte membrane, which is commercially known as Nafion®. The use of an anhydrous Nafion membrane in PEMFCs makes the membrane design more compact with high power density than other types of fuel cells. In addition to the electrolyte, PEMFCs use porous carbon electrodes with a high surface area coated with noble Pt as catalysts at the anode and cathode [6]. Like in the case of the AFC, in the PEMFC, pure H_2 and O_2 are used at their respective electrodes, which are supplied by the reformers. PEMFCs operate in a relatively low

Table 2.1 Main specifications of the various types of fuel cell technologies: AFC, PEMFC, PAFC, MCFC, DMFC and SOFC

Specifications	AFC	PEMFC	PAFC	MCFC	DMFC	SOFC
Electrolyte	Potassium hydroxide (KOH)	Perfluorosulfonic acid (PFSA)	Phosphoric acid	Alkaline carbonates of (Potassium, Sodium or Lithium)	Perfluorosulfonic acid (PFSA)	Metal oxide ceramics
Temperature (°C)	60–80	50–70	160–220	550–700	30–130	900–1000
Pressure (bar)	<30	1.3–7	0.7–0.9	1–8	1–3	0.7–1
Cathode	Ni/Ni-Mo	Pt/Pt–Pd	Pt/Pt–Pd	Ni	Pt/Pt–Pd	LSM-YSZ
Anode	NiO/Pt/Pt–Pd	RuO_2/IrO_2	RuO_2/IrO_2	Ni–Cr	Pt/Pt–Pd	Ni-YSZ/Ni-ceramic
Electrical efficiency (%)	40–60	50–60	37–45	40–50	35–50	40–50
Stack life (kh)	5–8	25–30	> 40	35–40	8–20	35–40
Applications	Aerospace	Transportation, small devices	Transportation, Distributed generation	Large consumers, Distributed generation	Portable devices	Power plants and Distributed generation
Capital cost ($/kW)	400–600	75	~3000	~3000	~2200	~1290

temperature range (50–70 °C), which allows them to start quickly without warming up and to have a dynamic response compared to AFCs. However, the presence of a proton-conducting electrolyte requires the use of a noble catalyst (i.e., Pt) on electrodes to split hydrogen into protons and electrons, which adds a significant cost to the overall system [7,8]. Additionally, Pt is very susceptible to CO poisoning, which degrades its electrocatalytic activity. Therefore, it is very important to purify the input hydrogen gas if it is derived from hydrocarbons [9]. Thus, the purification of H_2 requires a reactor, which adds additional cost to the full stack design. Currently, researchers around the world are exploring cost-effective and CO-resistant alloys of Pt and Ru as potential alternatives that can mitigate these challenges [10]. The compact design, high power density and dynamic response of PEMFCs allow them to be used in stationary and particularly transportation applications, such as hydrogen fuel cell electric vehicles (FCEVs). However, a major issue in using PEMFCs in vehicles is the storage of hydrogen (discussed earlier in section 3). High-energy-density fuels composed of long- and short-chain hydrocarbons can be used as precursors to produce hydrogen, but the system must be reformed for the conversion of methanol/ethanol to hydrogen gas. This not only increases the cost and maintenance of FCEVs but also pollutes the environment through the emission of GHGs. However, these emissions are lower than the current emissions from modern internal combustion engines [11].

2.3   Phosphoric Acid Fuel Cell

Phosphoric acid fuel cells (PAFCs) are among the most advanced commercially available hydrogen fuel cell technologies. The PAFC was developed in mid-1960 as a result of research on high-temperature hydrogen fuel cells [12]. PAFCs mainly differ from PEMFCs due to the use of different electrolytes. As shown in Figure 2.1c, the liquid phosphoric acid (PA) is embedded in a Teflon-bonded silicon carbide matrix, which acts as an electrolyte for the transfer of protons generated at the anode and high surface area porous carbon electrodes dispersed with a Pt catalyst. This enables PAFCs to operate at higher temperatures (160–220 °C), thus overcoming the shortcomings of PEMFCs, which can occur due to low-temperature operation, such as water management, high activation losses and sluggish kinetics [13]. In addition, the high-temperature operation also allows the use of hydrogen directly from the reformer, therefore eliminating the gas cooling step. Moreover, when PA is used as an electrolyte, it can tolerate CO_2 to a greater extent and CO to 1–3%, which is usually mixed in reformed hydrogen gas [14]. Moreover, PAFCs have been found to be very useful as stationary power generation devices in defence and military applications [15]. They are generally very efficient (up to 85% efficiency) in the cogeneration of electricity in combined heat and power (CHP) systems but are less efficient in standalone applications (37–45%) [16]. Moreover, the use of PAs in the matrix results in PAFCs being comparatively larger than PEMFCs with higher weight/volume ratios for the same power density. Additionally, the use of noble Pt catalyst agglomeration, carbon corrosion, electrolyte depletion and fuel starvation reduces the lifetime and consequently increases the overall maintenance cost in PAFCs [17]. Although PAFC systems have achieved a lifespan of 40,000 h, they still have not been widely commercialized. One of the primary reasons for this limited adoption is the substantial cost stemming from the above discussion. The main contributing factor to the high cost of PAFCs is the use of expensive noble Pt, which has agglomeration issues; therefore, frequent replacement is needed after a certain lifetime; for example, after 1000 h of operation at 175 mA/cm^2 and 175 °C, the agglomerate size of the Pt particles increased from 22 to 40 Å at the anode to 23 to 46 Å at the cathode [18]. This increase in agglomerate size directly correlates with the decrease in the surface area of the Pt electrocatalyst, which is the main reason for the degradation in PAFC performance.

2.4 Molten Carbonate Fuel Cell

The molten carbonate fuel cell (MCFC) was first introduced in the early 1950s by Broers and Katelaar [19]. MCFC operates in a similar manner to AFC, where a negative ion travels from the cathode anode through a charge-selective exchange membrane where it recombines with a proton created as a result of the hydrogen oxidation reaction (HOR). The operation of MCFC differs from that of AFC due to the different negative species that travel through the electrolyte. As shown in Figure 2.1d, unlike in AFCs, a mixture of both oxygen and carbon dioxide is supplied at the cathode, where it reacts with an electron that has travelled from the anode through an external circuit and formed a carbonate species (CO_3^{-2}). This carbonate ion travels toward the cathode, where hydrogen gas (supplied at the anode) is oxidized by carbonate ions to form water and carbon dioxide. MCFCs typically operate in the temperature range 550–700 °C to maintain sufficient ionic conductivity in the molten state [20]. This temperature range reduces the activation energy barrier and therefore prevents the use of noble metals as electrode materials for electrochemical oxidation and reduction processes [21]. During its early stages, the electrodes used in MCFC were precious metals; however, the technology evolved with time, and eventually, the noble metals were replaced with nickel-based alloys for anodes and nickel oxide as a cathode material [22]. The use of nonprecious metal electrocatalysts significantly reduces the cost compared to that of PEMC and PAFC. Moreover, the high electrical efficiency of MCFC compared to PAFC is another reason; for example, the stand-alone efficiency in MCFC is reported to reach 60%, which is almost double that of PAFC (37–42%). This reported efficiency in MCFC power generation can further increase up to 85% in CHP systems where waste heat is captured and used in MCFC plants [23]. These fuel cells were developed for large natural gas and coal-based power plants for industrial and commercial electric utilities. Another advantage of MCFC over PAFC is that high-temperature operation does not require an external reformer to convert dense long-chain hydrocarbons to hydrogen; rather, due to the high temperature, this energy-dense fuel is converted to hydrogen gas inside MCFCs through an internal reforming process, which also reduces the additional cost. In regard to CO/CO_2 poisoning, MCFCs possess a distinctive advantage compared to other fuel cells because of the carbonate electrolyte. In fact, they can utilize the CO_2 mixed in fuel, therefore allowing them to utilize the impure H_2 gas as a result of coal reforming. Despite their relative resistance to CO_x poisoning, researchers around the world are actively working to further enhance MCFC resilience against impurities found in coal, e.g., sulfur oxide, nitrous oxide and other particulates [24]. The main disadvantage of current MCFCs is durability. High-temperature operation together with the use of corrosive electrolytes causes severe corrosion of electrodes and other fuel cell components, which decreases cell life [25]. Scientists and engineers are currently exploring high-temperature corrosion-resistant materials for electrodes as well as fuel cell design to mitigate this problem without decreasing their efficiency.

2.5 Direct Methanol Fuel Cell

Unlike other hydrogen fuel cells, direct methanol fuel cells (DMFCs) use methanol (CH_3OH) mixed with steam instead of hydrogen as fuel at the anode (Figure 2.1e). PEMs are a subset of PEMFCs that have the same carrier (H^+) from the anode to the cathode [26]. Methanol and steam enter the anode where they are oxidized in the presence of catalysts, releasing H^+, electrons and CO_2. The protons pass through an ion exchange membrane (i.e., PFSA or Nafion®), while the electrons pass through an external circuit to the cathode where they recombine in the presence of oxygen to form water (the same as in a PEMFC). It is important to properly vent CO_2 outside the cell. DMFCs are a relatively new technology that are used in small-scale portable power applications with sizes ranging from 100 mW to 100 W [27] and operating at temperatures ranging from 30 to 130 °C [28]. The major advantage of DMFCs over other fuel cells is the nonrequirement of fuel storage since methanol is a high-energy density fuel and is easy to transport to consumers with current infrastructure [29]. Another advantage over other types of fuel cells is the high power density, which makes them competitive with Li-ion batteries for small-scale portable electronic devices. Scientists and researchers have encountered various challenges in DMFCs, which are similar to those of PEMFCs because the present DMFC has adopted the design of the current state-of-the-art PEMFC. These materials are composed of common components, such as the use of a Nafion electrolyte and Pt as electrode catalysts. However, these approaches have not only encountered the same problems as PEMFCs but also had several additional problems, such as sluggish methanol oxidation reactions on Pt and Ru catalyst surfaces; therefore, DMFCs require Pt loading almost ten times greater than that of PEMFCs to create the same number of H^+ ions from methanol oxidation, as is the case for H_2 gas in

PEMFCs. This adds a significant cost in stack design [30]. The use of Nafion in DMFCs causes severe methanol crossover (30–50%) problems toward the cathode because of the high miscibility with water. A high crossover temperature not only reduces the oxygen reduction reaction (ORR) efficiency at the cathode due to the intolerance of the Pt electrocatalyst to methanol but also reduces the concentration of both fuel and water at the anode, which reduces the overall performance [31]. In addition, methanol crossover also causes significant water flooding at the cathode. Researchers are modifying membranes to reduce methanol crossover by treating surfaces with different additives and designing membranes with new materials that enhance ionic conductivity while discouraging fuel crossover [32].

2.6 Solid Oxide Fuel Cell

Solid oxide fuel cells (SOFCs) are solid-state ionic conducting electrochemical energy generators that use nonporous ceramic compounds as electrolytes for the transfer of oxygen ions. As shown in Fig. 2.1f, oxygen enters the cathode side where it reacts with the electron that has transferred from the anode though an external circuit and forms an oxide ion (O^{2-}). O^{2-} travels through the solid electrolyte toward the anode, where it combines with H_2 fuel and forms water while releasing electrons. SOFCs operate at very high temperatures (900–1000 °C), which enables the internal reforming of fuels at the stack [33]. Additionally, high-temperature operation reduces the energy barrier for catalysts; hence, an SOFC removes noble catalysts [34]. The solid electrolyte in SOFCs allows flexible configurations (tubular, planar, and monolithic designs) and greater tolerance to sulfur and CO poisoning than other fuel cells, which allows them to use gases from coal gasification [35]. The high-temperature operation and internal reforming of SOFCs make them 40–50% efficient at converting fuels to electricity [31]. Conversely, this approach also has several drawbacks, e.g., a slow startup time, effective thermal shielding to maintain high temperature and personnel safety. Therefore, this approach is feasible for utility applications but impractical for transportation and portable use. In addition, a high operating temperature also demands durability for the use of materials in SOFCs. Currently, the major research challenge in SOFC technology is the development of cost-effective electrodes and conductive electrolyte materials that are highly durable at practical cell operating temperatures; e.g., ceramics of Yttria-stabilized zirconia (YSZ) and rare earth element-doped ceria are currently the most studied materials for high-temperature SOFC applications [36]. Additionally, several other problems are the need for high-temperature bonded sealing materials for gastight stacking and the lack of durable interconnect metal materials for intermediate-temperature (650–800 °C) operations [31]. All the major technical specifications are summarized in Table 2.1.

If hydrogen is supplied to the fuel cell and the membrane electrode assembly is stable, this technology will operate indefinitely. However, the high price of clean hydrogen, incompatible hydrogen infrastructure and technological maturity hinder the commercialization of fuel cells [37–39]. To reduce the overall cost of fuel cell technology, some major issues that should be addressed include replacing noble electrode catalyst materials with cost-effective transition metal-based alloys, designing inexpensive solid electrolyte membranes and increasing green hydrogen production. In addition, an economical and easily accessible supply of hydrogen is imperative for fuel cells to compete with existing technologies in the commercial market.

Chapter 3
Computational Fluid Dynamics

Computational Fluid Dynamics (CFD) is a branch of fluid mechanics that is concerned with numerical solution of differential equations governing transport of mass, momentum, and energy in moving fluids [40]. It involves using computer algorithms and simulations to solve complex fluid flow and heat transfer phenomena problems, which may be difficult or impossible to solve analytically.

CFD activity emerged and gained prominence with availability of computers in the early 1960s, and since early 1970s commercial software packages became available making CFD an important component of engineering practice in diverse areas [40]. Nowadays, the role of CFD in engineering predictions has become so strong that today it can be viewed as a new 'third dimension' in fluid dynamics, the other two dimensions being the classical cases of pure experiment and pure theory [41].

The fundamental principle behind CFD is based on the conservation laws of mass, momentum, and energy. These laws, known as the Navier–Stokes equations, describe the behavior of fluids.

CFD comprises the creation of computer programs and their subsequent testing, following a numerical methodology that involves different key steps: problem definition, pre-processing, discretization, solution, post-processing and validation.

- Problem definition, identifying the physical domain geometry of study and the objectives.
- Pre-processing: the problem domain geometry, the mesh generation and the physics are defined, subdividing the complete domain into certain number of control volumes.
- Discretization: the governing partial differential equations and the boundary conditions of the problem are converted respectively into linear algebraic equations and boundary grid-points. There are three types of discretization methods: Finite Difference Method (FDM), Finite Volume Method (FVM), and Finite Element Method (FEM), being FVM the most commonly used and widely adopted in commercial CFD software packages due to its versatility and ability to handle complex geometries and conservation laws.

A. Iranzo et al., *Computational Fluid Dynamics Modelling of PEM Fuel Cells*, SpringerBriefs in Energy, https://doi.org/10.1007/978-3-032-06631-2_3

- Solution: the discretized linear algebraic equations are solved iteratively using numerical algorithms until convergence is achieved. The computational domain is divided into time steps, and at each time step, the equations are solved for the flow variables within each control volume. Residuals, representing the imbalance between of the discretized equations and calculated as the difference between the left-hand side and the right-hand side, play important roles in assessing the accuracy and convergence of the solution. As the solution converges, the residuals should decrease until they reach a desired level of tolerance.
- Post-processing: once the solution is obtained, results are analysed and interpreted. This step involves visualizing the flow field, extracting relevant data, and analyzing parameters such as velocity profiles, pressure distributions, temperature distributions, and forces acting on objects in the flow.
- Validation: A CFD code or software developed is finally tested by setting up and running the code for certain problems (called benchmark problems), for which accurate numerical or experimental results are available in the published literature. Testing the accuracy of a computer simulation as compared to published analytical/numerical results is called verification whereas the comparison with the experimental results is called validation of a CFD code. The former test verifies the correctness of the programming and computational implementation of a mathematical/computational model and latter confirms the agreement of the computer simulation with the physical reality; within certain acceptable level of inaccuracy.

If the obtained results do not accomplish the defined the simulation goals, the model must be revised and updated until achieving the objectives (Fig. 3.1).

CFD has become an essential tool in engineering and scientific research, as it allows engineers and scientists to study and predict the behavior of fluid flows in a wide range of multiple applications. These applications include diverse engineering areas such as energy systems, aerospace, automobile, turbomachinery, chemical, electrical, electronics, biomedical and energy systems among many others [42].

In the field of energy systems, CFD enable engineers to optimize their designs and improve overall efficiency, providing valuable insights into diverse energy research areas such as: Oil and Gas, Combustion and Gasification, Turbomachinery, Renewable Energy, Fuel Cells and Hydrogen or Nuclear Energy [43]. CFD activity in these energy systems areas has been demonstrated over the past few years through a significant number of scientific publications, as depicted in Fig. 3.2.

It can be observed that the Oil and Gas industry represents the major contributor to energy engineering in CFD research, studying diverse subjects such as liquid loading phenomena in gas wells, piping systems or liquefied natural gas production to mention some of them. CFD is playing also a major role in the aerodynamic design of Turbomachinery, and currently all modern designs of gas and steam turbines, hydraulic turbines and pumps among others are being aided by the use of CFD, with clear reductions in the costs and design cycles. Combustion and Gasification is as well one of the major fields of application of CFD, where the wide variety of combustion types have been deeply explored, including coal and biomass, liquids, gas, oxy-combustion, and others.

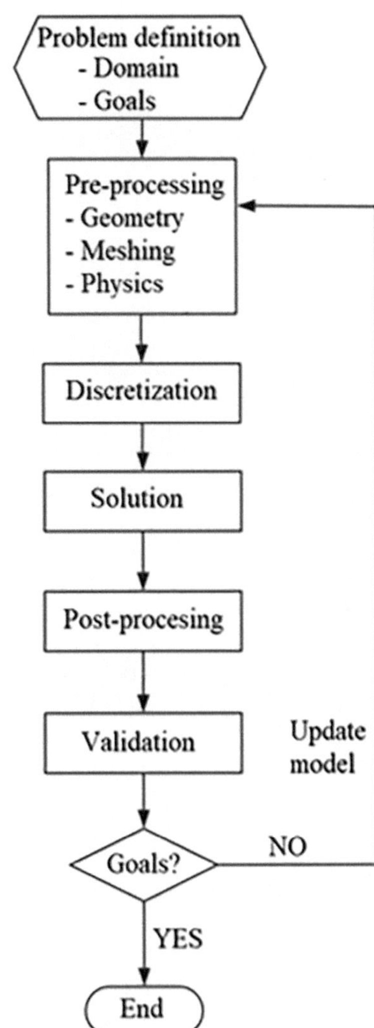

Fig. 3.1 CFD workflow diagram

Renewable energy represents an important subject, and has been also widely investigated with CFD, including wind energy, solar energy, ocean energy and biofuels. In wind energy and solar energy, major research efforts have been made respectively for enhancing turbine blade aerodynamics and for optimizing the heat transfer in solar collectors, receivers, solar air heaters among other components. In ocean energy, extensive research has been conducted in the optimization of wave energy converters and tidal stream turbines.

Finally, regarding biofuels and nuclear power generation, significant studies have been conducted related to biofuel production and safety analysis respectively. Extensive research has been also conducted in the area of fuel cells and hydrogen, where

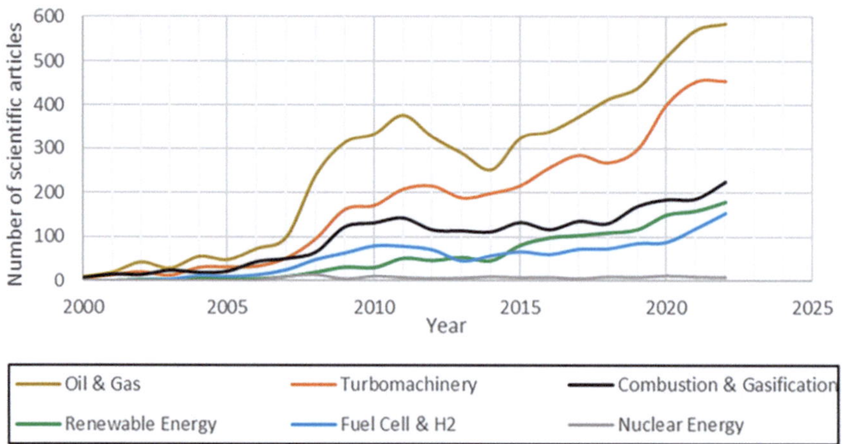

Fig. 3.2 Annual publications of scientific articles from 2000 to date in different areas of energy systems. *Source* Scopus

CFD is playing also a major role in the fuel cell design due to the capacity of modelling fluid flow, heat and mass transfer, and also electrochemistry, in order to compute the reactants oxidation and reduction rates.

Chapter 4
Computational Fluid Dynamics Modelling of PEM Fuel Cells

The operation of a fuel cell is the result of complex interactions among species transport, fluid flow, porous media, heat transfer, electrochemical reactions, potential fields, and electric current transport. This interrelation and coupling of various physical phenomena make the overall design of a fuel cell a complex process.

Due to the difficult experimental environment of fuel cell systems [44], many efforts have been stimulated to develop software tools using CFD technics, capable of solving numerically such complex fluid flow and heat transfer phenomena problems. CFD modelling permits analyzing the behavior of the fuel cell in order to define design criteria and optimal operating conditions, enabling the understanding of the processes taking place. The past few decades have seen the number of published modelling and simulation work on PEM fuel cells increased dramatically. This has essentially been made possible by recent advances in computational techniques and the availability of high-performance computing resources [45].

The different CFD fuel cell modeling strategies can be classified according to different criteria, such as the dimensionality of the computational domain, the scope of the computational domain or the assumptions and simplifications.

– Regarding the dimensionality of the computational domain, the pioneering CFD models were 1D considered majority the direction orthogonal to the membrane, with varying degrees of complexity and details [45]. The posterior 2D CFD models treated also the transport phenomena that occur inside the fuel cell in another additional direction, usually either the dimension along the GFC or that perpendicular to it. Although the pioneering 1D models and subsequent 2D models have significantly contributed to the advancement of PEM fuel cell modelling, these models can only be used for particular phenomena in PEM fuel cells due to the 3D nature of the transport processes occurring in a PEM fuel cell. Consequently, they are not suitable for parametric studies or optimization of fuel cell designs. Moreover, the transports of mass, heat, and charges in porous structures are major issues for these models. To accurately predict cell performance, a realistic 3D description of the cell geometry is necessary. Thus, 3D computational models have been developed to better

A. Iranzo et al., *Computational Fluid Dynamics Modelling of PEM Fuel Cells*,
SpringerBriefs in Energy, https://doi.org/10.1007/978-3-032-06631-2_4

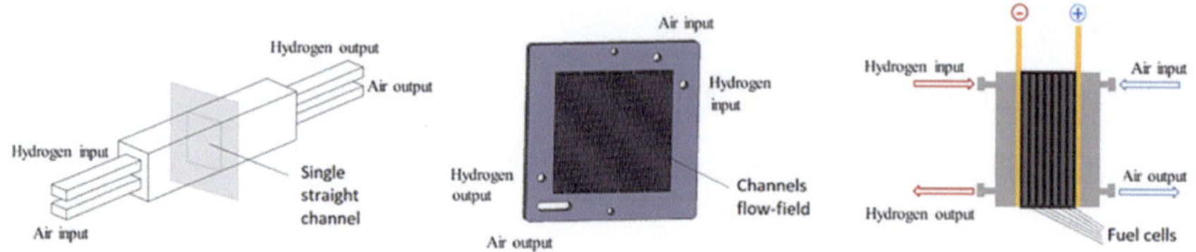

Fig. 4.1 Scope of the computational domain: straight channel, single cell, stack

understand the phenomena occurring within PEM fuel cells and then predict their performance though a full 3D description of a complete cell or a fuel cell stack requires a considerable computational effort. Three-dimensional models consider effects in both across and along the gas flow channel, in addition to the direction normal to the membrane. The bulk of the models treat phenomena such as water and reactant transport, electrochemical reactions, species and charge transport, mass transfer, heat transfer occurring during the cell operation, with the effects of geometry or operating conditions. Thus, depending on the characteristics of the phenomena being investigated, and the level of details included, the 3D models can be divided into three distinct groups: models for transport phenomena, models for geometry or operating condition effects, and models for thermal effects.

– Regarding the scope of the computational domain, CFD models can be referred to specific sections of the fuel cell, such as a single straight channel or the flow fields in the channels [46], a single fuel cell including all of its components [47], or a complete fuel cell stack [48], see Fig. 4.1. In some CFD models, it is preferred to limit the computational domain to only one straight flow channel. However, this type of simplification is not valid to analyze the effect of the flow field characteristics on the PEM fuel cell performance and water management, and thus, larger computational domains including the whole fuel cell or stack geometry are required for further and more detailed studies allowing for the resolution of flow characteristics and variables of interest such as concentrations, temperatures, current densities, etc. throughout the computational domain. Nevertheless, due to the high computational costs involved, computational domains are usually applied to single cells much more frequently than to complete stacks.

Regarding the assumptions and simplifications, regardless of the type of multi-phase flow model being used, it is often necessary to make some modelling assumptions in order to simplify the problem under consideration and thereby facilitate its numerical implementation without reducing the accuracy of the model. Some commonly used assumptions in PEM fuel cell modelling are summarized next:

– Steady state,
– Ideal gases,
– Laminar and incompressible gas flow,
– Local thermal equilibrium,
– Isotropic and homogeneous components materials,
– Uniform GDL compression.

Understanding the intricate physics, validating predictions, and utilizing specialized tools are key pillars in the advancement of Proton Exchange Membrane Fuel Cell (PEMFC) technology. These elements play a crucial role in ensuring the accuracy, reliability, and practical application of computational fluid dynamics (CFD) simulations in the context of PEMFCs, and are discussed in the following sections.

4.1 CFD Models for PEMFC Physics

The functioning of a fuel cell arises from complex interactions involving species transport, flow of fluids, permeable materials, heat exchange, electrochemical reactions, electric current transport, and potential fields. These interconnected relationships and interactions of diverse physical phenomena render the holistic design of a fuel cell a complex process.

In the following, the equations governing the physical and electrochemical processes involved in the PEMFC simulations are enumerated [46, 49].

4.1.1 *Conservation Laws of Mass, Momentum, and Energy (Navier–Stokes Equations)*

The fundamental principle behind CFD is based on the conservation laws of mass, momentum, and energy. These laws, known as the Navier–Stokes equations, describe the behaviour of fluids and can be expressed in its general form as the following Eq. (4.1) [50]:

$$\frac{\partial}{\partial t} \int_v \rho \emptyset dV + \oint_A \rho \emptyset V.dA = \oint_A \nabla \emptyset.dA + \int_V S_\emptyset dV \tag{4.1}$$

where ϕ is the transported variable, t is time, A is the surface area, V is the volume, Γ is the diffusivity of the variable ϕ, and S is the source of ϕ. The first term in the equation represents the transient transport of ϕ, the second term represents the transport by convection, the third term represents the transport of ϕ by diffusion, and the fourth term represents the source (or sink) of ϕ.

4.1.2 *Electrochemistry Model*

The exchange current densities in anode and cathode have the following general definitions (4.2), (4.3) [49]:

$$R_a = \left[\zeta_a j_a(T)\right]\left(\frac{[A]}{[A]_{ref}}\right)^{\gamma_a}\left(e^{-\frac{\alpha_a^a F \eta_a}{RT}} - e^{-\frac{\alpha_c^a F \eta_a}{RT}}\right) \tag{4.2}$$

$$R_c = \left[\zeta_c j_c(T)\right]\left(\frac{[C]}{[C]_{ref}}\right)^{\gamma_c}\left(e^{-\frac{\alpha_a^c F \eta_c}{RT}} + e^{-\frac{\alpha_c^c F \eta_c}{RT}}\right) \tag{4.3}$$

where j(T) is the reference exchange current density per active surface area, ζ the specific active surface area, [A|C] and [A|C]$_{ref}$ the respective local and reference species concentration, η is the surface overpotential, γ the concentration dependence, α^y_x the (dimensionless) charge transfer coefficients from the electrode x to the y, F the Faraday constant (9.65×10^7 C/kmol), R the universal gas constant, and T the temperature.

The exchange current densities j(T) depends on the local temperature in the form of Eqs. (4.4) and (4.5) [49]:

$$j_c(T) = j_c^{ref}\, e^{\frac{E_c}{RT}\left(1-\frac{T}{T_c^{ref}}\right)} \tag{4.4}$$

$$j_a(T) = j_a^{ref}\, e^{\frac{E_a}{RT}\left(1-\frac{T}{T_a^{ref}}\right)} \tag{4.5}$$

with E the reversal potential given by Nernst equation, and j^{ref} the reference exchange current density at the given temperature T^{ref}.

4.1.3 Dissolved Phase Model

The generation and transport of dissolved water is described as Eq. (4.6) [49]:

$$\frac{\partial}{\partial t}\left(\frac{\varepsilon_i M_{H2O}\rho_i}{EW}\lambda\right) + \nabla\left(\frac{j_m n_d}{F}M_w\right) = \nabla\left(M_w D_w^i \nabla\lambda\right) + s_\lambda + s_{gd} + s_{ld} \tag{4.6}$$

where M_{H2O} is the molecular weight of water, ε is the porosity of the porous media and j_m is the ionic current density vector given by the Eq. (4.7):

where λ is the dissolved water content, $D_w{}^i$ the diffusion coefficient of water content, and S_λ, S_{gd} and S_{ld}, are the rates of water generation in CL, of mass change between gas and dissolved phases, and of liquid and dissolved phases, respectively.

$$j_m = -\sigma_{mem}\nabla\varnothing_{mem} \tag{4.7}$$

The S_{gd} and S_{ld} rates are given by Eqs. (4.8), (4.9) [49]:

$$S_{gd} = \frac{\left(1 - S^\theta\right)\gamma_{gd}M_{H2O}\rho_i}{EW}(\lambda_{eq} - \lambda) \tag{4.8}$$

$$S_{ld} = \frac{S^{\theta}\gamma_{ld}M_{H2O}\rho_i}{EW}(\lambda_{eq} - \lambda)$$ (4.9)

with ρ_i the dry membrane density, EW the equivalent membrane weight, s the liquid saturation level, λ_{eq} the equilibrium water content, γ_{gd} and γ_{ld} the gas and liquid mass exchange rate constants, and θ the exponent that is used to compute the phase-change rates.

The equilibrium water content is computed at the same time [32] as Eq. (4.10) [49]:

$$\lambda_{eq} = 0.3 + 6a[1 - \tanh(a - 0.5)]$$
$$+0.69(9.2 - 3.52)a^{1/2}\left[1 + \tanh\left(a - \frac{0.89}{0.23}\right)\right] + s(16.13 - 9.2)$$ (4.10)

where the water activity a is given by Eq. (4.11) [49]:

$$a = \frac{p_{ww}}{p_{sat}}$$ (4.11)

were p_{wv} and p_{sat} are the water vapor partial pressure and the saturation pressure, respectively.

4.1.4 *Liquid-Phase Modelling*

The transport equation for the liquid phase is driven by the pressure gradient V_{pl} following [32] as Eq. (4.12) [49]:

$$\frac{\partial}{\partial t}(\varepsilon_i \rho_i S) = \nabla\left(\frac{\rho_l K K_r}{\mu_l}\right)\nabla p_l + S_{gl} - S_{ld}$$ (4.12)

where ρ_l is the liquid density, μ_l the liquid dynamic viscosity, K and K_r the absolute and relative permeability, and p_l the liquid pressure. The relative permeability is computed in the GDL as Eq. (4.13):

$$K_r = S^b$$ (4.13)

where s is liquid saturation, b is a given exponent, and in membrane can be expressed as Eq. (4.14):

$$K_r = \left(\frac{\left(\frac{M_{H2O}}{\rho_l}\right)\lambda_{s=1} + \frac{EW}{\rho_i}\lambda}{\frac{M_{H2O}}{\rho_l}\lambda + \frac{EW}{\rho_i}\lambda_{s=1}}\right)^2$$ (4.14)

Because the liquid pressure can be expressed as the sum of the capillary pc and gas pressure p, Eq. (4.12) can be rewritten as Eq. (4.15):

$$\frac{\partial}{\partial t}(\epsilon_i \rho_l s) = \nabla\left(\frac{\rho_l K K_r}{\mu_l}\nabla(p_c + p)\right) + S_{gl} - S_{ld} \qquad (4.15)$$

For the calculation of the capillary pressure, Leverett function is applied. In addition, liquid water reduce the effective active surface are in CLs, so that the transfer currents are penalized by Eq. (4.16):

$$R_j = (1 - s)^{\gamma_j} R_j \qquad (4.16)$$

where γ_j is the exponent accounting for the liquid blockage to the reaction surface.

4.1.5 Properties

The equations for computing the properties of the gas phase species diffusivity, the diffusivity of water content and the osmotic drag coefficient are described next [49].

– Gas phase species diffusivity.

Gas phase species diffusivities can be computed using the dilute approximation method expressed by Eq. (4.17):

$$D_i = \varepsilon^{1.5}(1 - s)^{r_s} D_i^0\left(\frac{p_0}{p}\right)^{\gamma_p}\left(\frac{T}{T_0}\right)^{\gamma_T} \qquad (4.17)$$

D^0_i being the mass diffusivity of species i at the reference temperature and pressure (p_0, T_0) [33], and r_s the exponent of pore blockage. In general, $p_0 = 101{,}325$ Pa, $T_0 = 300$ K, $\gamma_p = 1.5$ and $\gamma_T = 1$.

– Diffusivity of water content.

The diffusivity coefficient in the water content (dissolved phase) transport equation is computing using Eq. (4.18):

$$D_w^i = \frac{n_\lambda \rho_i}{EW}f(\lambda) \qquad (4.18)$$

where n_λ is a generic coefficient, and the function f(λ) has been computed from Wu et al. [34] as Eq. (4.19):

$$f(\lambda) = 4.1e^{-10}\left(\frac{\lambda}{25}\right)^{0.15}\left[1 + \tanh\left(\frac{\lambda - 2.5}{1.4}\right)\right] \qquad (4.19)$$

– Osmotic drag coefficient.

The osmotic drag coefficient can be calculated using Eq. (4.20):

$$n_d = n_{osm} g(\lambda) \tag{4.20}$$

where η_{osm} a generic factor with the default value of 1.0 and the function $g(\lambda)$ is given by Eq. (4.21):

$$g(\lambda) = \frac{2.5\lambda}{22} \tag{4.21}$$

4.1.6 Multi-phase Flow in PEMFC Modelling

Multiphase flow plays a vital role in predicting the fluid behavior with real life applications in the field of academia and industry processes. The complexity of the two-phase model increases due to considerations of interface regions between catalyst layer and membrane to understand interaction mechanism within the scope of PEMFC where the exchange of mass, momentum and heat are also taken into account unlike the single-phase flow simulations. The flows as shown in Fig. 4.2 are majorly classified as into dispersed consists of droplets and bubbly flow, while the intermittent flow is of slug and churn and separated flow consist of film, annular horizontal stratified flows. To the specific application of these in the occurrence of PEMFC gas channels, found to have multi-phase flow regimes accounting with a differential in pressure drop throughout the gas channels. For instance, in mist flows when the liquid droplets are dispersed into a fine small droplets with high gas velocities where there is no extensive effects on the pressure drop and closely agrees to a single-phase flow as low amount of water content in gas channels, but as with increase of water content with gas velocity a droplet flow regime can be observed with varying the pressure drop trend. This trend can be further transformed to a thin film of liquid along the channel walls. With lower gas velocities and increased water content can build up the slug flow.

The CFD modelling of PEMFCs requires not only the resolution of the regular CFD equations (conservation of mass, momentum, species, and energy). In addition, the CFD model must account for the equations for the conservation of charge, the electrochemistry modelling, the multi-phase flow in channels and porous media including water transport. These addon modules increases the additional challenges to resolution of a PEM fuel cell.

Multiphase modeling approaches are categorized mainly into two, the Eulerian-Eulerian approach and Eulerian–Lagrangian models [52]. In Eulerian- Eulerian approach, the different phases are interpreting continuous while having the volume fraction of each phase. Lagrangian approaches rely on tracking the dispersed phase,

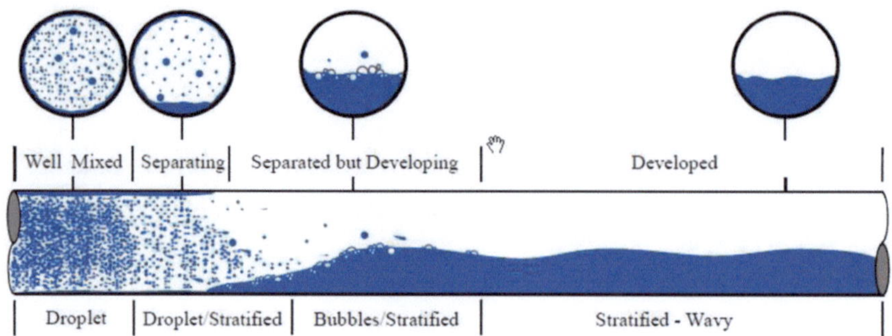

Fig. 4.2 Flow regimes in the flow channel [51]

consisting of parcels of particles (Lagrangian) which move through the computational fluid domain, while interacting with the continuous bulk phase (Eulerian).

The Volume of Fluid (VOF) model is widely utilized in two phase flow simulation of PEM fuel cell, especially tracking the interaction of gas–liquid interface in gas channels and can be easily incorporated along the effects of surface tension and wall adhesion. In two-phase flow, continuity and Navier stokes equation for the gas/liquid mixture is solved. [53].

Equation of continuity:

$$\frac{\partial \rho}{\partial t} + \nabla.(\rho v) = S_m \tag{4.22}$$

Here, ρ is the mixture density, and v the velocity vector of the mixture.

Equation of momentum (Navier Stokes equation):

$$\frac{\partial}{\partial t}(\rho v) + \nabla.(\rho vv) = -\nabla P + \nabla.\tau + \rho g + F \tag{4.23}$$

Here, p is the pressure shared by both phases, g and fs represents the volumetric gravitational and surface tension forces, respectively. where τ is a viscous tensor, In Newtonian fluid, which is often assumed in fluid dynamics simulations, the stress tensor is linearly related to the velocity gradient by the dynamic viscosity.

$$\tau = \mu\left(\nabla v + (\nabla v)^T\right) \tag{4.24}$$

The mixture density and viscosity are calculated as sum of the phase (air/water) properties.

$$\rho_m = \alpha_{air}\rho_{air} + \alpha_{water}\rho_{water} \tag{4.25}$$

$$\mu_m = \alpha_{air}\mu_{air} + \alpha_{water}\mu_{water} \tag{4.26}$$

In Gas channels, the force related to surface tension (F) from the equation of momentum can be calculated as

$$F = -\sigma K \nabla \alpha_i \tag{4.27}$$

$$K = \nabla . n \tag{4.28}$$

where the terms σ represents the surface tension, $\nabla \alpha_i$ is the volume fraction gradient normal to the interface and K is the curvature of phase interface accounting in VOF models. The wall adjacent normal vectors of interface are adjusted according to the equation.

$$n = n_1 . Cos\theta + n_2 . Sin\theta \tag{4.29}$$

where n1 and n2 denote the unit vectors in the normal and tangential directions, θ is the contact phase angle.

Mixture Multiphase (MMP) Approach

The MMP is a simplified Eulerian- Eulerian approach, used as an efficient method to investigate multiphase problems. In this approach a single equation for the two-phase phenomena is solved while only changing the sources terms of PEMFC components thus rendering it simple, unified and consistent approach. This method generally assumes that the two phases are miscible and at equilibrium, and that their motion can be simulated as that of a unique continuum. Similarly to the VOF approach, in MPP conservations of mass and momentum equations are solved.

$$\frac{\partial \rho}{\partial t} + \nabla . (\rho v) = S_m \tag{4.30}$$

The momentum equation of the mixture can be expressed as

$$\frac{\partial}{\partial t}(\rho v) + \nabla . (\rho v v) = -\nabla P + \nabla . \tau + \rho g + F + \nabla . (\sum \alpha_i \rho_i u_{dr}) \tag{4.31}$$

where u_{dr} is the drift velocity and can be written as $u_{dr} = u_l - u_m$

Eulerian Multiphase (EMP) Model

Water management plays a significant role in enhancing fuel cell performance such that a balance between the water retention and removal capability is crucial. In addition, the balance effectively minimizes ohmic losses by humidified membrane and also removal of water content within the porous electrodes, gas channels minimize the mass transport losses. The water transport mechanism in porous electrodes is mainly characterized by the operational conditions of fuel cell which are driven by capillary pressure, phase and Knudsen diffusion with the porous media while the gas channel transport mechanism is classified based on the flow regimes.

$$\frac{\partial}{\partial t}(\alpha_i \rho_i v_i) + \nabla.(\alpha_i \rho_i v_i v_i) = -\alpha_i \nabla P + \alpha_i \rho_i g - \alpha_i \gamma_i v_i - \alpha_i \nabla P_{ci} \qquad (4.32)$$

where α_i corresponds to the volume fraction of species i, ρ_i is the density of species i, v_i the velocity of species i, P is the pressure, g the gravitational acceleration, γ_i and P_{ci} are the viscous and capillary pressure forces. As with the viscous force dominance, the momentum equation can be written as follows.

$$-\nabla P - \gamma_i v_i = 0 \qquad (4.33)$$

In PEFC, the body forces such as gravity are often neglected. However, for the transport mechanism in porous material, the momentum governing equation is not explicitly used whereas the empirical multiphase Darcy law is employed.

$$\gamma_i = \frac{\mu}{K_i.\alpha_i^n} \qquad (4.34)$$

When the above viscous relation is substituted in the momentum equation as

$$\nabla P = -\frac{\mu}{K_i.\alpha_i^n} V_i \qquad (4.35)$$

where K is the intrinsic permeability of porous media in phase i.

The capillary pressure due to the difference between these gaseous and liquid phases is given as

$$P_c = P_g - P_l \qquad (4.36)$$

$$P_c = \sigma \, Cos\theta \left(\frac{\varepsilon}{K}\right)^{0.5} J(s) \qquad (4.37)$$

$$J(s) = \begin{cases} 1.42(1-s) - 2.12(1-s)^2 + 1.26(1-s)^3 \theta < 90 \\[2mm] 1.42s - 2.12s^2 + 1.26s^3 \theta > 90 \end{cases} \qquad (4.38)$$

where P_c is the capillary pressure at the interface of two different porous electrode is continuous varying the contact angle, porosity and permeability due to discontinuous liquid saturation.

In source terms in the gas channel and solid parts are generally neglected while for the porous media (GDL, CL) the viscous and inertial contribution to flow resistance are expressed using Darcy Law. Figure 4.3 shows the transients simulation performed on a single differential fuel cell. The initial condition used in this single channel represents the steady state full humidification. During these conditions the liquid water exists in both anode and cathode. Upon changing to dry gasses in both the compartments, liquid water level under the channels gets evaporated and removed by supplied inlet gasses. This can be termed as the dewetting characterization of GDL through-plane direction [54]. Moreover, the under-channel experience significant drop in water content showing the effects on performance than the ribs of the channels. Due to high water diffusivity from anode, the drying of anode GDL is faster than that of the cathode side.

Multiphase in Porous Media

Let us start by looking at a schematic picture of the processes that occur in the gas diffusion layer (GDL), the catalytic layer (CL), and the membrane electrolyte in a hydrogen proton exchange membrane fuel cell (PEMFC). The GDL and CL form what is called a gas diffusion electrode (GDE). Both the GDL and the CL contain electron conducting carbon and gas pores. In addition, the CL contains electrochemically active catalyst particles, as well as pore electrolyte pressed into the pores of the electrode.

To minimize interface resistances, it is also common to cover the side of the GDL facing the CL with a microporous layer (MPL). The MPL has similar properties as the GDL, but with a finer pore structure. From a modeling perspective, an MPL can be handled as a GDL.

At the anode, hydrogen is transported through the gas pores in the GDL and then through a thin layer of pore electrolyte (green) in the CL to finally be adsorbed at an active site in the catalyst. Then, hydrogen dissociates and is oxidized to form hydrogen ions (protons) by releasing its electrons. The electrons are conducted out

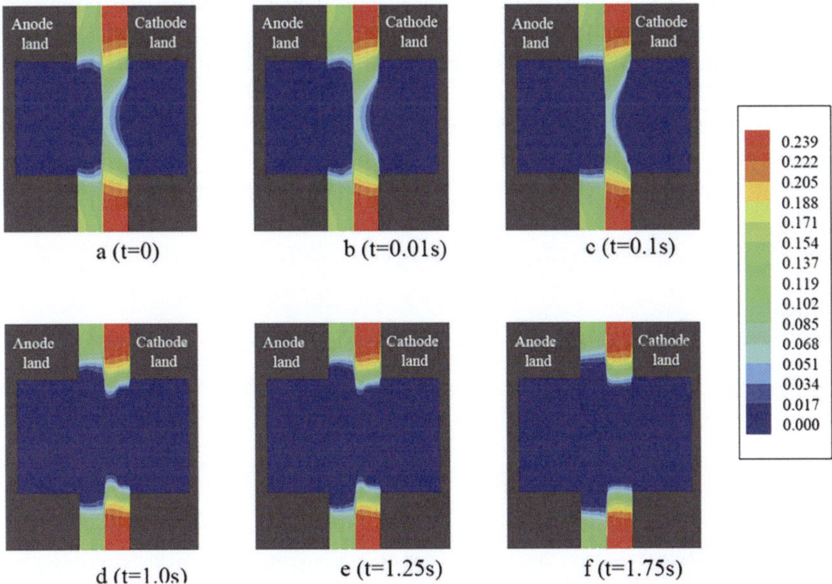

Fig. 4.3 Transient simulations of liquid water volume fraction under anode and cathode channel [54]

of the electrode through the electronic-conducting electrode material to the current collector, which conducts the electron through the external circuit to the cathode. The total sum reaction for the charge transfer reaction at the anode is:

$$H_2 = 2H^+ + 2e^- \qquad (4.39)$$

At the cathode, oxygen is transported through the GDL and then through a thin layer of pore electrolyte to the active sites in the catalyst. Oxygen adsorbs, dissociates, and is reduced by the electrons that have been conducted through the external circuit and through the electrode to the active site. Here, oxygen also takes up protons to form water. The total sum reaction is thus:

$$O_2 + 4H^+ + 4e^- = 2H_2O \qquad (4.40)$$

It can be observed in the charge transfer reactions in Eqs. 4.1 and 4.2 that two hydrogen molecules are required to generate the electrons needed to reduce one molecule of oxygen in Eq. 4.40.

In the pore electrolyte and in the membrane electrolyte, protons formed at the anode migrate to the cathode. Water may be dragged along with the protons. However, since water is also formed at the cathode, its higher activity here may lead to back diffusion from the cathode to the anode.

Figure 4.5 shows the principles for homogenization of the domains depicted in Fig. 4.4. The upper sketch of Fig. 4.5 depicts a heterogenous representation of the GDLs and the CLs. The electrode particles (light blue and orange), gas pores, and pore electrolyte occupy their own respective positions in space. Where there is pore electrolyte, there are no particles. Where there are particles, there are no gas pores, etc.

In the lower sketch of Fig. 4.5, the GDLs and the CLs are represented as slab mixtures. The GDLs are represented as slab mixtures of the gas pores and electrode particles while the CLs are represented as slab mixtures of the solid electrode

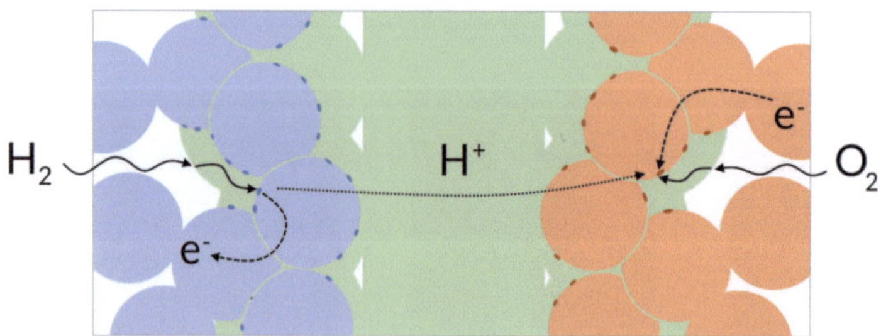

Fig. 4.4 A schematic drawing of the processes that occur inside the electrodes and electrolyte in a PEMFC. Light blue represents the anode (negative electrode) while orange represents the cathode (positive electrode). The membrane electrolyte and the pore electrolyte (the electrolyte in the CL) are represented in light green. *Credit* COMSOL AB

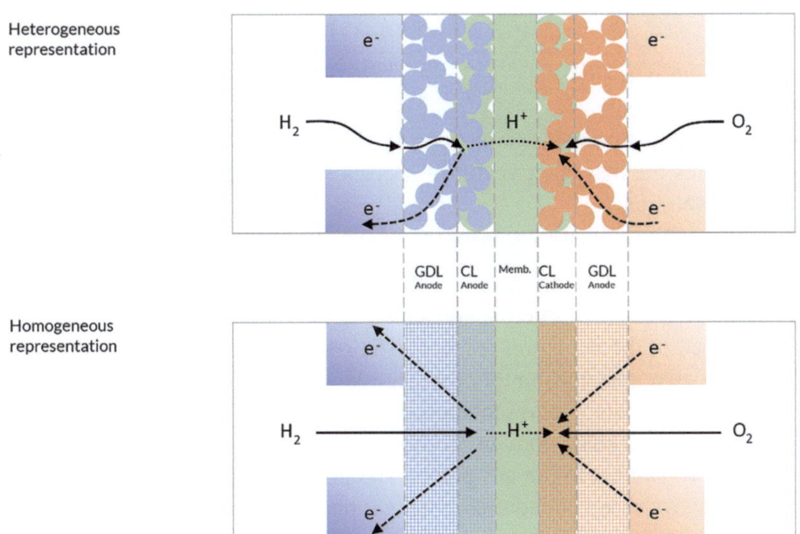

Fig. 4.5 Heterogeneous (top) and homogeneous representations (bottom) of the GDLs and CLs in the PEMFC gas diffusion electrodes (GDEs). *Credit* COMSOL AB

particles, gas pores, and pore electrolyte. The gas pores, electrode particles, and pore electrolyte (CL only) coincide at the same points in space [55]. The GDL is described using a porosity, which gives the fraction of electrode particles and gas pores at each point in space in the slab. In the same fashion, the CL is described by the respective volume fractions of the electrode particles, gas pores, and pore electrolyte in each point in the domain.

The Balance and Transport Equations

Considering both the heterogeneous and homogeneous representations in Fig. 4.5, both representations require the same number of balance equations.

In order to understand the behavior of the cell it is needed to solve the material balances for all species in the gas phase on both the anode and cathode sides. A material balance can be formulated for each species in the gas phase except one, whose concentration is derived from a mass balance formulated for the sum of all species, i.e., the total flux balance.

For a homogeneous representation, the mass balance for all species except one might look like:

$$\frac{\partial \left(\varepsilon_p \rho \omega_i \right)}{\partial t} + \nabla \cdot \left(\boldsymbol{j}_i + \rho \omega_i \boldsymbol{u} \right) = R_{i,tot} \tag{4.41}$$

where:

- ε_p denotes porosity
- ρ denotes density
- ω_i denotes the mass fraction of species i
- $\mathbf{j}_i$ denotes the mass flux vector of species i relative to the velocity vector of the whole solution
- $\mathbf{u}$ denotes the velocity vector of the whole gas phase solution

$R_{i,tot}$ denotes all of the reactions where species i is involved. For example, in the respective CL, hydrogen is consumed in the oxidation reaction, while oxygen is consumed in the reduction reaction. The reaction term, $R_{i,tot,}$ may thus represent these respective reactions in the two electrodes.

In a heterogeneous representation, the porosity in Eq. (4.41) is 1 since there is only gas in the gas phase domain. Correspondingly, there are no reaction terms unless oxygen is consumed in the gas phase. The phase boundary of the pore electrolyte and the electrodes are boundaries to the gas phase domain. Boundary conditions can be applied here instead of the reaction terms of the homogeneous representation. For example, at the gas–pore electrolyte boundary, it can be assumed that the flux of hydrogen across this boundary is equal on both sides. In the following equations a homogeneous representation will be assumed.

In Eq. (4.41), the mass flux vector can be expressed implicitly, as given by the following expression:

$$-\nabla x_i + \frac{1}{p}(x_i - \omega_i)\nabla p = -\sum_{j \neq i} \frac{x_i x_j}{D_{ij}} \left(\frac{\mathbf{j}_i}{\omega_i \rho} - \frac{\mathbf{j}_j}{\omega_j \rho} \right), \qquad (4.42)$$

where:

- x_i denotes the mole fraction of species i
- p denotes the absolute pressure
- D_{ij} denotes the binary diffusivity for the pair of species ij

Note that this would be an effective diffusivity for the porous medium in the homogeneous representation and could also include Knudsen terms for the collision of molecules with the pore walls. In Equation (4.42) [56], the right-hand side (RHS) represents the friction between species i and j as they move in the solution. The difference in flux between the two species leads to friction forces, which are proportional to this difference. The left-hand side (LHS) represents the driving force of the flux of species i. Equation 4.42 states that the friction of species i with all other species in the solution (RHS), as it moves through the solution, must be balanced by the driving force, which is the gradient in chemical potential (LHS). The relation between the mole fraction and the mass fraction is given by the relation:

$$x_i = \frac{\omega_i}{M_i} \left(\sum_j \frac{\omega_j}{M_j} \right)^{-1}, \qquad (4.43)$$

where M_i denotes the molar mass for species i. Note that the inverse expression within parentheses denotes the mean molar mass:

$$M_n = \left(\sum_j \frac{\omega_j}{M_j} \right)^{-1} \tag{4.44}$$

The last equation in the material balances is the mass balance of the solution. Since the mass flux vector is relative to the movement of the whole solution, the total sum of the mass fluxes must be zero:

$$\sum_i j_i = 0 \tag{4.45}$$

One material balance equation is obtained per species except one, which can be obtained using Eq. (4.45). In addition, there is an equation per mass flux vector, since Eq. (4.42) is formulated for every species in the solution. Equations (4.41)–(4.45) provide one equation per unknown, except for pressure, p, density, ρ, the velocity vector, $\boldsymbol{u}$, and the overpotential that governs the electrochemical reactions in the term $R_{i,tot}$ in Eq. (4.41). When accounting for multiphase flow also porosity would be a dependent variable, which would be obtained from a continuity equation for multiphase flow.

The relationship between pressure and density provides an additional equation. This relationship may be represented by the ideal gas law or some other equation of state. The ideal gas yields the following expression:

$$\rho = \frac{p}{RT} M_n \tag{4.46}$$

Note though that this introduces temperature, T. (R denotes the universal gas constant). If the fuel cell is not running under isothermal conditions, then it is required to add an equation for temperature. Normally the ideal gas law is used, although other equations of state are available.

The fluid flow equations for the whole solution yield the velocity and pressure fields. For porous media flow, Darcy's law or the Brinkman equations can be used to define the continuity and momentum equations. For example, the Brinkman equations can be expressed as:

$$\frac{\partial(\varepsilon_p\rho)}{\partial t} + \nabla \cdot (\rho\boldsymbol{u}) = Q_m$$

$$\frac{\rho}{\varepsilon_p}\left(\frac{\partial\boldsymbol{u}}{\partial t} + \boldsymbol{u} \cdot \left(\frac{1}{\varepsilon_p}\nabla\boldsymbol{u}\right)\right)$$

$$= \nabla \cdot \left(\frac{1}{\varepsilon_p}\left(\mu(\nabla\boldsymbol{u} + (\nabla\boldsymbol{u})^T) - \frac{2}{3}\mu(\nabla \cdot \boldsymbol{u})\boldsymbol{I}\right)\right) \quad\quad (4.47)$$

$$- \nabla p - \left(\frac{\mu}{\kappa} + \frac{Q_m}{\varepsilon_p^2}\right)\boldsymbol{u} + \boldsymbol{F}$$

In this system of equations, Q_m denotes a mass source due to: the electrochemical reactions (for example, hydrogen forms protons that are not present in the gas phase), condensation, or evaporation (if several phases are modeled). Furthermore, μ denotes viscosity, k denotes the permeability of the porous medium, and $\boldsymbol{F}$ denotes a volume force vector, for example, buoyancy. Also, in the case where only single-phase flow is modeled in the GDEs, water formed in the cathode reaction may evaporate into the gas pores and, in this way, give a mass source in the continuity equation and corresponding material balance for water, which would also yield the volume reaction of water and thus also the porosity as water fills the gas pores. (The reaction picks up four protons from the pore electrolyte per oxygen molecule consumed as it forms one water molecule per oxygen atom.)

For free (open) media (for example, in the gas channels), the system in Eq. (4.47) is replaced by the Navier–Stokes equations. Advanced CFD PEMFC solvers are able to model multiphase flow in porous media as well as in free media. In such cases, the phase transport equations that utilize a multiphase flow mixture model are added to the system of equations. If the flow is turbulent, additional equations are added depending on the turbulence model selected. A large number of Reynolds-averaged Navier–Stokes (RANS) turbulence models (for single-phase and multiphase flow) as well as large-eddy-simulation (LES) and detached-eddy-simulation (DES) capabilities can be available.

With the preceding system of equations, there are now an equal number of equations as unknowns, except for temperature and the overvoltage in the reaction terms in Eq. 4.41. The next model equation is the energy equation:

$$\rho C_p\frac{\partial T}{\partial t} - \nabla \cdot (k\nabla T) + \rho C_p\boldsymbol{u} \cdot \nabla T = \left(\frac{\partial p}{\partial t} + \boldsymbol{u} \cdot \nabla p\right) + Q, \quad\quad (4.48)$$

where:

- C_p denotes the heat capacity
- k denotes the effective thermal conductivity
- Q is a generic heat source term

The heat source term can include Joule heating, entropy effects, activation losses from the electrochemical reactions, heat of condensation, and viscous heating (which is often neglected). The pressure work term (the first term on the RHS) is often neglected in fuel cells because the pressure gradients are usually relatively small.

Equations 4.41–4.48 yield an equal number of relations as unknowns, except for the overpotential in the electrochemical reactions included in Eq. 4.41 and in the mass sources or sinks in the continuity equation in the system described in Eq. 4.47.

Considering the reaction term, it can be expressed as:

$$R_{i,tot} = \sum_m R_{i,m},$$

(4.49)

where $R_{i,m}$ is the mth reaction in which species i is involved. For simplicity, it can be assumed that there is only one electrochemical reaction and that no other reactions occur in the GDE. Then, in the catalytic layer, we have:

$$R_i = \frac{M_i \nu_i a_\nu}{nF} i_{loc},$$

(4.50)

where:

- ν_i denotes the stoichiometric coefficient of species i in the electrochemical reaction
- n denotes the number of electrons in the electrochemical reaction
- a_ν denotes the specific surface area of the catalyst (unit catalyst surface area per unit volume)
- i_{loc} denotes the local current density at the surface of the catalyst

The local current density is given by a Butler–Volmer expression:

$$i_{loc} = i_{0,ref} \left(\prod_{i:\nu_i>0} \left(\frac{p_i}{p_{i,ref}} \right)^{\nu_i} exp\left(\frac{\alpha_a F \eta}{RT} \right) - \prod_{i:\nu_i<0} \left(\frac{p_i}{p_{i,ref}} \right)^{-\nu_i} exp\left(-\frac{\alpha_c F \eta}{RT} \right) \right)$$

(4.51)

In this expression:

- $i_{0,ref}$ denotes the exchange current density at the reference conditions (where the equilibrium potential was measured)
- p_i denotes the partial pressure of species i
- $p_{i,ref}$ denotes the partial pressure of species i at the reference conditions
- α_a denotes the anodic charge transfer coefficient
- η denotes the overpotential of the electrochemical reactions
- α_c denotes the cathodic charge transfer coefficient

The partial pressure of species i is obtained from the mole fraction and total pressure, so this does not introduce a new unknown. The overpotential is a new unknown. This can be expressed as:

$$\eta = \phi_s - \phi_l - E_{eq},$$

(4.52)

where:

- ϕ_l denotes the local electric potential in the electrolyte (ionic conductor)
- ϕ_s denotes the local electric potential in the electrode (electrode particles, i.e., the electronic conductor)
- E_{eq} denotes the equilibrium potential at the reference conditions. That is: the difference between the electrode and electrolyte potentials at a zero current density, when the cell is operating at the reference pressures

With Eq. 4.52 there are two new unknowns: the local electrode potential and local electrolyte potential. The two last equations we need are the equations for the balance of current. For the PEMFC models, Ohm's law can be utilized for both the electrolyte and the electrode. In the pore electrolyte in the CL, the following expression gives the ionic current balance:

$$\nabla \cdot (-\sigma_l \nabla \phi_l) = a_v i_{loc} \tag{4.53}$$

where σ_l denotes the conductivity of the electrolyte. In the membrane electrolyte (outside the pore electrolyte), the RHS in Eq. 4.53 is equal to zero, as there are no electrochemical reactions in the membrane electrolyte outside the pore electrolyte. This equation is the simplest form for the balance of current in the electrolyte. If electroosmotic water drag is activated, an additional contribution to the ionic current is added, which is proportional to the gradient in water activity. The water activity is then obtained by solving a separate material balance for water in the electrolyte. If an impedance spectroscopy or transient analysis is selected, then a source of current is added on the RHS (in addition to i_{loc}), which accounts for the charge and discharge of the double layer. This double-layer term is the product of the time derivative of the overpotential multiplied by the volumetric double layer capacitance.

The last equation is the balance of current in the electrode. This is formulated in the catalytic layer as:

$$\nabla \cdot (-\sigma_s \nabla \phi_s) = -a_v i_{loc}, \tag{4.54}$$

where σ_s denotes the conductivity of the electrode. Again, we could also add the contribution of the double layer capacitance here. In the GDL, the RHS is zero.

With Eqs. 4.41–4.54, there is the same number of unknowns and equations that can be solved with the numerical model using the finite or volume element method.

4.2 CFD Model Validation

Model validation is a fundamental stage in PEM Fuel Cell CFD modelling. Modelling equations described in the previous section contains several model parameters. While many of them can be clearly identified in supplier datasheets (such as thermal or electrical conductivities) there are still many parameters which are unknown and must be experimentally determined, or used as fitting parameters during model validation. Parameters that can be determined experimentally are for example the interfacial contact resistance between Bipolar Plate and GDL, or catalyst layer porosity. Among the most challenging parameters to be determined are the ones related to the electrochemistry of the electrodes. There are experimental techniques to determine them [57] but typically at least the reference exchange current density (io_{ref}) will be complex to be determined with accuracy. It means that this parameter is in many cases used as a model fitting parameter when experimental data is available for the fuel cell being modelled.

However, it must be considered that a comprehensive validation in CFD PEM fuel cell modelling is significantly challenging due to the complex and highly-coupled phenomena occurring within the cell, and the high number of model parameters in the equations being solved. In practice, this means that model validation against a unique polarization curve is not guaranteeing at all that a model is properly and accurately validated (see the excellent review by Wang [58]). Instead, a set of minimum three polarization curves obtained for different operating conditions should be used for validation, and also of high importance is that if possible two different flow field designs are used for validation. The reason for this is that both the Membrane-Electrode Assembly (MEA) and the flow-field design are significantly influencing the cell performance (polarization curve) but the physical (and electrochemical models in the MEA) are significantly different as indicated in Sect. 4, so that validating a CFD model with different flow-field designs and operating conditions is providing a more comprehensive and ensuring the quality and trust of the CFD model (see for example [59]).

Needless to say, all geometrical details, material thicknesses and properties, operating conditions, must be the same between the CFD model and the experimental item being used for validation. It is sometimes observed in publications that a model validation is carried out against experimental data from the literature, where indeed the experimental data is corresponding to a different fuel cell (flow field design, MEA or GDL properties) than the model intended to be validated, which is obviously not a valid procedure.

It is also well recommended to provide additional value to the validation process by using if possible local distributions of variables, such as current density, temperature, or water distributions. This is obviously significantly more complex in terms of experimental facilities than obtaining polarization curves, but if the experimental data is available either by means of Current Density Mapping [60], water visualization techniques such as Neutron Imaging [61] or others [62], this is providing a meaningful and coherent validation of the CFD model being developed.

As an example, Figs. 4.6 and 4.7 presents results of a CFD model for a 50 cm^2 cell, together with experimental data measuring liquid water within the *in-operando* fuel cell obtained by means of Neutron Imaging [63, 64]. The CFD software used was ANSYS Fluent [63] and AVL FIRE M [64].

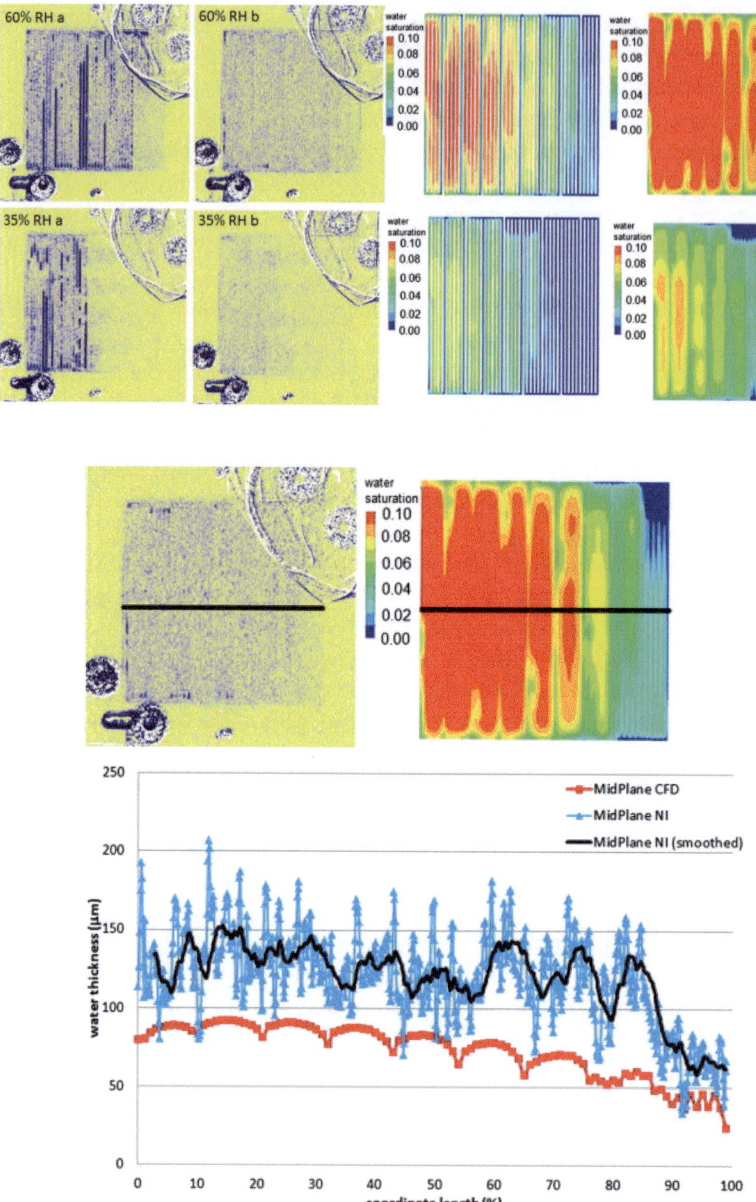

Fig. 4.6   Top: neutron radiographs obtained at Paul Scherrer Institut for a 50 cm^2 fuel cell during operation (left) and after flushing the liquid water within the channels (middle-left). CFD simulations (ANSYS-Fluent) showing the water saturation at the GDL-Bipolar Plate interface and GDL-Catalyst Layer interface [63]. Bottom: Quantitative comparison of the mid-plane distribution of liquid water saturation [64]

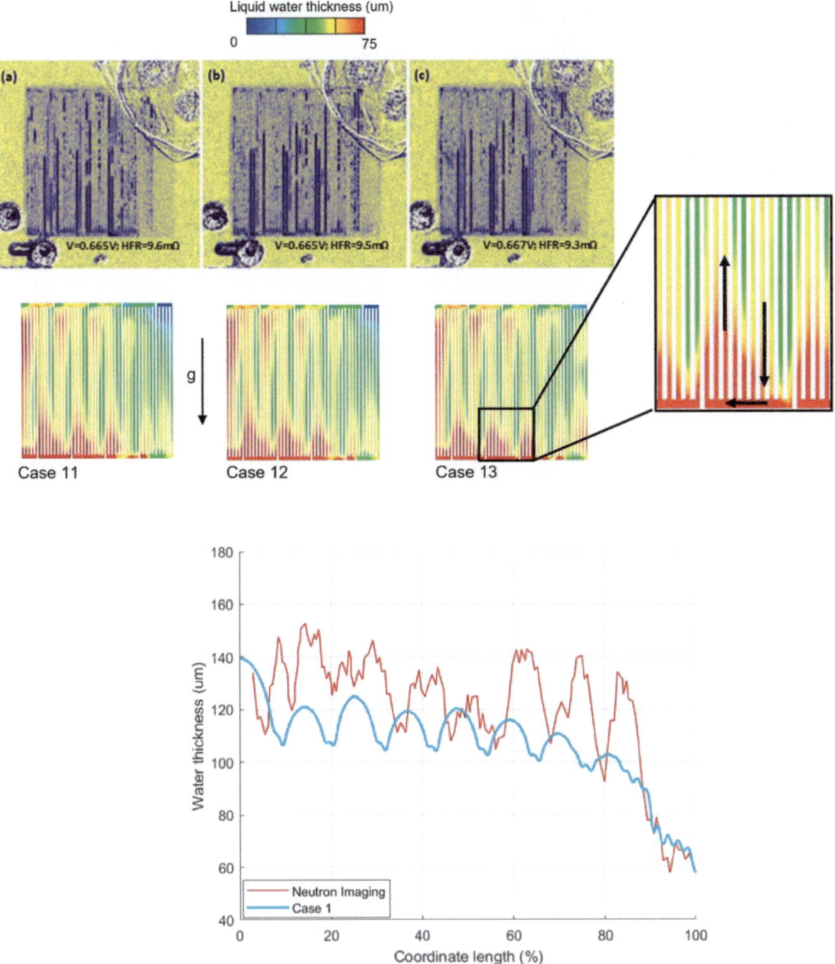

Fig. 4.7 Top: neutron radiographs obtained at Paul Scherrer Institut for a 50 cm^2 fuel cell during operation and corresponding CFD simulations (AVL FIRE M) showing the water saturation at the GDL-Bipolar Plate interface [64]. Bottom: Quantitative comparison of the mid-plane distribution of liquid water saturation [64]

4.3 Commercial Software for PEM Fuel Cell Modelling

In this section the most widely used commercial software for PEM Fuel Cell modelling is addressed, and unique features are highlighted thanks to the kind contribution from the software vendors.

4.3.1 Software AVL FIRE™ M

The PEM fuel cell model of AVL FIRE M [65] has been described and compared to measurements in many publications [66–75]. This section contains the highlights and outstanding features of the solution approach.

Meshing

The generation of suitable computational meshes for industrial fuel cell designs is a major challenge. The computational meshes generated with AVL FIRE M for the fuel cell application consist of a combination of polyhedral elements and extruded prismatic elements. The latter is used for the meshing of thin regions, such as the MEA, and wherever possible to reduce the mesh size and enhance the result quality. AVL FIRE M offers the possibility to automatically generate such extrusions on several parts: in- and outlets, flow channels in the flow direction and porous transport layers (catalyst layer, microporous layer, gas diffusion layer, membrane) in the cell normal direction. Usually, the automatic meshing process treats cathode and anode separately, creates extrusions in cell normal direction and connects the cathode and the anode in the middle of the membrane via a non-conformal interface. Figure 4.8 shows two examples with automatically extruded prism elements.

If a single cell out of a stack is calculated, the user often faces the difficulty of defining realistic boundary conditions at the cut positions. Here, periodic boundary conditions can be defined during the meshing procedure to couple the flow and heat transport from one side to the other. In Fig. 4.8, periodic boundary conditions for the cooling channels are schematically shown. Without periodic boundaries, it would be necessary to include the adjacent bipolar plates, which would increase the cell count dramatically.

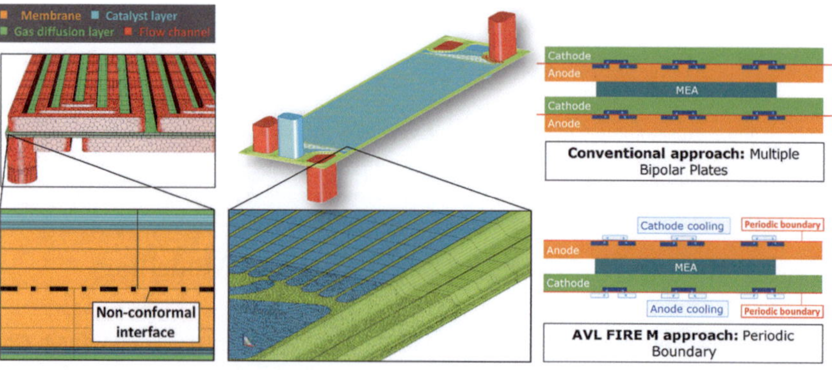

Fig. 4.8 Left and center: automated extrusion in cell normal direction and along flow channels; right: coolant flow with periodic boundaries

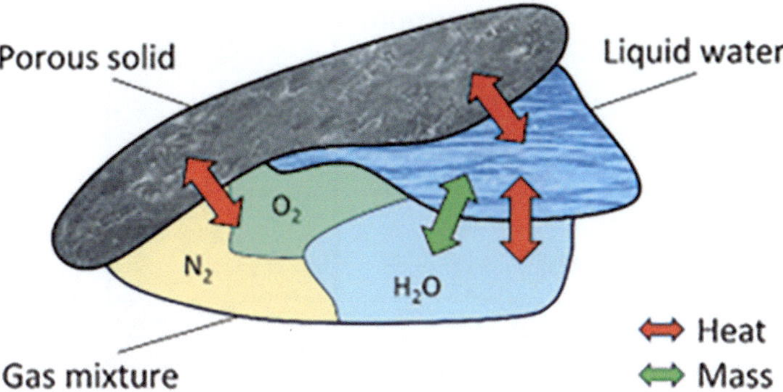

Fig. 4.9 Heat and mass transfer in a porous transport layer

Multiphase Modeling

The multiphase model for PEM fuel cells is based on the Eulerian approach where each fluid phase possesses individual velocities, volume fractions and temperatures. In the porous transport layers (catalyst layer, microporous layer, gas diffusion layer), three main phases coexist: a gas mixture, a liquid and a porous solid. For the gas mixture and the liquid, the intrinsic phase velocities are solved from momentum equations in which the interaction with the porous walls (concept of relative permeability) and the capillary forces are included. As shown in Fig. 4.9, mass transfer is considered between the fluid phases and heat transfer between all three phases.

At the interface between gas diffusion layer (GDL) and flow channel, the water droplets exiting the GDL are dragged towards the channel outlet by the gas phase. Depending on the geometry and the gas flow velocity, a portion of the liquid water remains stuck on the GDL modifying the interface capillary pressure and, thus, the liquid water flow inside the GDL. To include this effect in the model, the interface capillary pressure between flow channel and GDL is explicitly taken into account.

Catalyst Layer Modeling

The structure of the catalyst layer is described with a flooded agglomerate model. Here, the catalyst layer is assumed to consist of spherical agglomerates of platinum dispersed carbon particles embedded in ionomer covered by a thin ionomer film—see Fig. 4.10. The model describes the diffusion of the reactants (O_2, H_2) across the ionomer film into the agglomerate where the electrochemical reaction is solved locally depending on the local reactant concentration. Compared to the macro-homogeneous catalyst layer model in which no structure is assumed, the agglomerate model yields more realistic results at higher current densities, i.e. stronger voltage losses, due to the higher diffusion resistance. The agglomerate model is coupled with the degradation model (see below) in which agglomerate radius and ionomer film thickness are modified due to the volume changes of carbon, platinum and ionomer. In the catalyst layer, water exists in three phases: gaseous, liquid and dissolved in

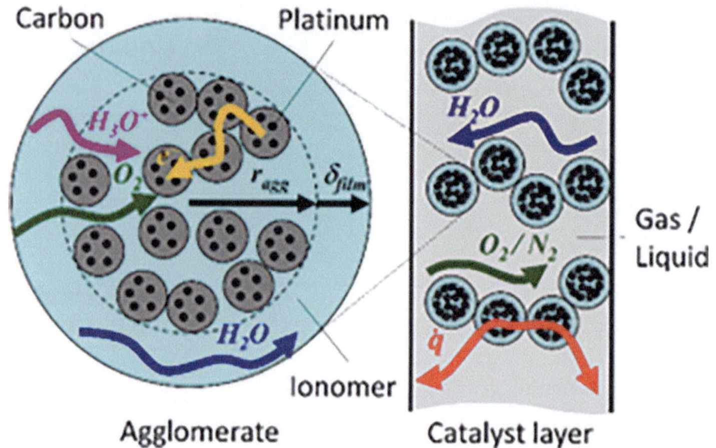

Fig. 4.10 Agglomerate model with transport quantities

the ionomer. Mass transfer between all phases (adsorption, desorption, evaporation, condensation) is properly included.

Membrane Modeling

In the membrane the 3D transport of dissolved water, ions and heat is solved. For the transport of water, three transport mechanisms are considered: diffusion, migration (electro-osmotic drag) and convection. The last mechanism is caused by a hydraulic pressure difference between the electrodes. In addition, the crossover of each gas species (O_2, H_2, N_2, …) dissolved in the ionomer phase is calculated. The parasitic redox reactions at the opposite electrodes triggered by the crossover of O_2 and H_2 are calculated as well, resulting in an accurate prediction of the real open circuit voltage. This means that the open circuit voltage is calculated rather than prescribed by the user.

Degradation Model

The degradation model focusses on chemical degradation of catalyst layers and membrane. The following degradation effects are covered:

- Carbon support corrosion triggered by carbon and platinum oxidation
- Platinum dissolution with redeposition
- Particle detachment with agglomeration
- Chemical ionomer degradation

An important result of the first three degradation effects is the reduction of the electrochemically active surface area (ECSA) due to changes in the particle size distribution and Pt oxidation. The ionomer degradation model calculates the production of hydrogen fluoride, thinning of the membrane and reduction of the ionic conductivity.

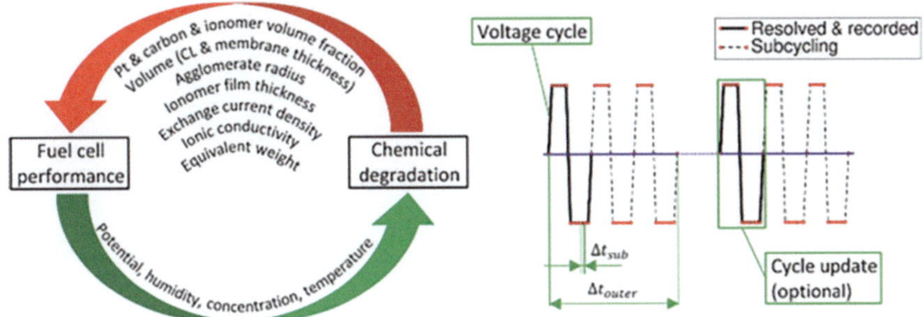

Fig. 4.11 Left: interdependency between fuel cell performance and chemical degradation; right: sketch of sub-cycling method

The degradation model is fully coupled with the performance model. Figure 4.11 left shows the feedback loop between fuel cell performance and chemical degradation. Spatial and temporal distributions of stressors (potential, humidity, concentration, temperature) are calculated with the performance model and provided as input for the degradation model. In the degradation model, the reaction equations for the individual degradation mechanisms are solved and material parameters (volume fractions, exchange current density, ionic conductivity, …) are modified influencing the fuel cell performance. With the new material parameters new values for the stressors are calculated, and so on.

A major challenge in the implementation of degradation models into a 3D CFD framework, is the simultaneous existence of two largely different time scales: The degradation mechanisms responsible for the aging of a fuel cell occur on a large time scale (hours, days or months). However, the degradation inducing operating conditions (usually voltage cycles) vary on a small time-scale (a few seconds). In 3D CFD calculations, in addition to the temporal also spatial details are resolved, usually in all three dimensions Resolving all spatial and temporal details would cause the computation time to increase dramatically. In the presented model this problem is overcome by applying a sub-cycling method for periodically repeating load profiles (e.g. voltage cycles). The main idea is, to record the local stressors of the first voltage cycle(s) and apply them in the background during a large outer time step while the transport equations are frozen—see Fig. 4.11 right.

Homogenized Channel Approach

For applications where a full resolution of the flow channels would be too expensive (ultra-fine structured flow fields or fuel cell stacks), AVL FIRE M offers the possibility to replace the flow channels with a porous medium, thus reducing the mesh size by a factor of 10–50. Although not all details are resolved, e.g. the regions below land and channel cannot be separated anymore, the overall result accuracy is comparable to the method with fully resolved channels—see Fig. 4.12.

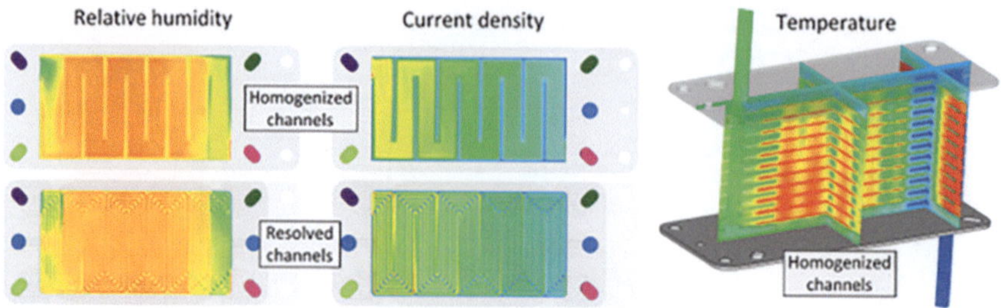

Fig. 4.12 Left: comparison between homogenized channel approach and fully resolved channels; right: stack simulation with homogenized channels

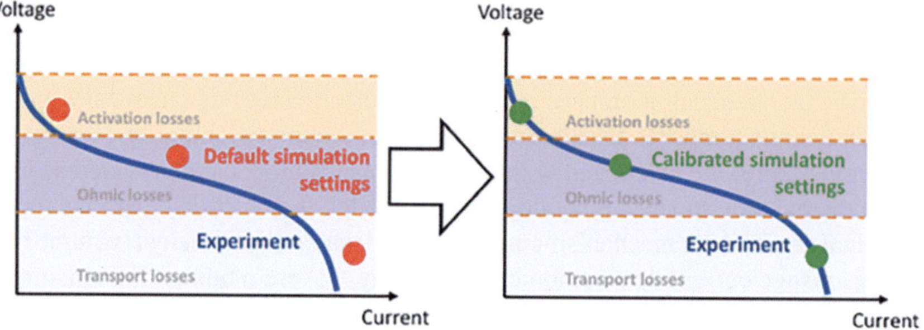

Fig. 4.13 Automatic calibration of simulation settings to an experimental polarization curve

Automatic Calibration of Fuel Cell Model

The fuel cell model can be calibrated automatically based on an experimental polarization curve without the necessity for geometrical simplifications. For each region of the polarization curve (activation, ohmic, transport) a specific set of calibration parameters is defined in a certain validity range. During the calibration process, simultaneously running simulation cases exchange values of calibration parameters resulting in a unified parameter set. With the newly found values of the calibration parameters, the complete experimental polarization curve can be calculated with high accuracy—see Fig. 4.13.

4.3.2   Software ANSYS

PEM fuel cell involves highly coupled physical phenomena such as fluid flow, heat transfer, diffusion of species, electric and ionic conduction, and multiphase transport. This physics is tightly coupled, hence any simplification while capturing the physics could cause inaccuracies in the results. In addition to this, the fuel cell designs consist of complex bipolar plate designs making it challenging from a pre-processing standpoint. Keeping all these intricacies in view, Ansys tools have been developed to address all these challenging requirements on the solver and pre-processing side. Ansys Discovery enables users to create and/or manipulate the complex geometries of fuel cells and prepare the computational domain, while Ansys Fluent provides state-of-the-art meshing and solver capabilities. The Fuel Cell Module in Ansys Fluent captures detailed physics relevant to fuel cells thus enabling users to generate accurate results which match closely with experimental data [76].

Computational Domain

In Ansys Fluent, the components of the fuel cells resolved as a part of the domain are:

• Current collectors	• Gas channels	• Gas Diffusion Layer (GDL)	• Membrane
• Coolant channels	• Catalysts	• Micro Porous Layer (MPL)	

The membrane electrode assembly (MEA) is modeled as three separate volumes—one each for catalyst on anode and cathode side, and one for the membrane. The GDLs, MPLs, catalysts and membrane are modeled as porous zones. The internal porous structure of the MEA is not captured in the geometry, but its effects are incorporated through various sub-models (Fig. 4.14).

Pre-Processing

Typical fuel cell bipolar plate designs include straight or serpentine shaped channels [77]. Depending on the manufacturing process of the plates (formed graphite or stamped metallic), the cross-section shape of channels can vary as shown in Fig. 4.15 [78]. The square/rectangular cross sections are relatively easier to capture in mesh compared to the trapezoidal shapes of the stamped metallic plates. Ansys Fluent Meshing offers a variety of meshing methods to effectively mesh various cross sections. The multi-zone meshing method can be used to generate a swept mesh with hex-dominant meshing. For complex shapes, polyhedral elements along with prism layers offer less laborious and robust meshing alternative (Fig. 4.16).

While the bipolar plate and the channel designs could be complex for meshing, the GDL/MPL, Catalyst layers and the membranes are typically rectangular shaped. A hex mesh with orthogonal quality $= 1$ can be generated in these layers easily using the multi-zone approach, while the complex bipolar plates and channels can be meshed using polyhedral elements. The two sets of meshes can then be connected using a non-conformal interface as shown in Fig. 4.16. For stamped metal plates, polyhedral mesh can be generated in the fluid and the solid metal plates as well as shown in Fig. 4.17.

Solver

Ansys Fluent models several transport equations to accurately account for the transport of mass, momentum, species, and energy. A dual potential equation is solved for the protonic and electronic potentials. Figure 4.18 below lists all the equations that are solved and regions in which these equations are solved [79].

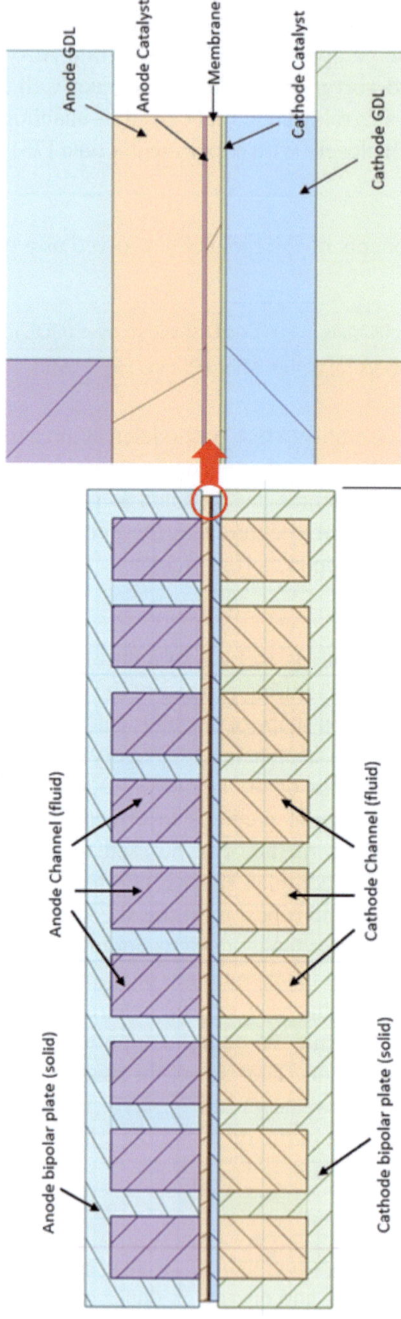

Fig. 4.14 Zones included in computational domain

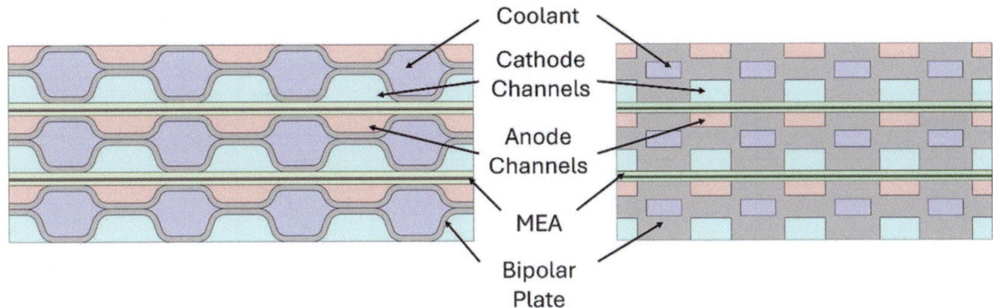

Fig. 4.15 Schematic of a cross sectional view of a stamped metallic (left) and a graphite (right) plate

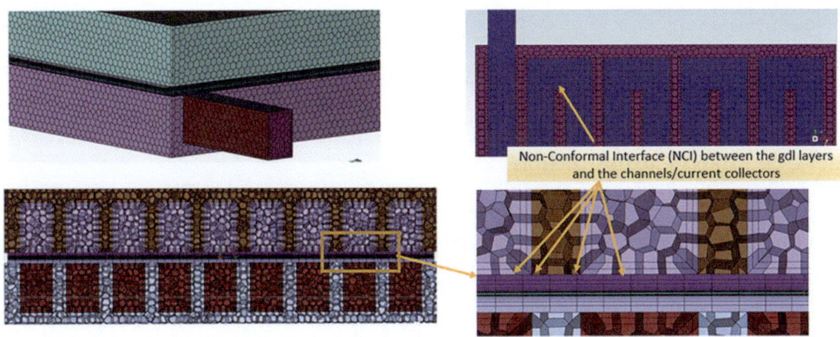

Fig. 4.16 Non-conformal mesh

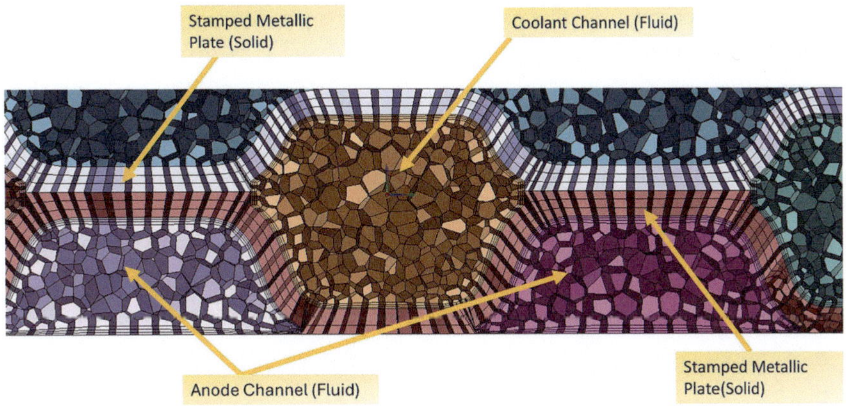

Fig. 4.17 Mesh in a stamped metallic plate

	Governing Equations	Flow	Energy	Species			Potential		Water Management		
				H_2	O_2	H_2O	Solid Phase (Electronic)	Membrane (Protonic)	Water Transport in Channels	Dissolved Phase Model	Water Transport in Electrodes and Membrane
	Region										
Anode	Cooling Channel (Fluid)	✓	✓								
	Current Collector (Solid)		✓				✓				
	Gas (H_2) Channel (Fluid)	✓	✓	✓		✓			✓		
	Gas Diffusion Layer (GDL)	✓	✓	✓		✓	✓				✓
	Micro Porous Layer (MPL)	✓	✓	✓		✓	✓				✓
	Catalyst Layer (Fluid)	✓	✓	✓		✓	✓	✓		✓	✓
	Membrane (Solid)		✓					✓		✓	✓
Cathode	Catalyst Layer (Fluid)	✓	✓		✓	✓	✓	✓		✓	✓
	Micro Porous Layer (MPL)	✓	✓		✓	✓	✓				✓
	Gas Diffusion Layer (GDL)	✓	✓		✓	✓	✓				✓
	Gas (O_2) Channel (Fluid)	✓	✓		✓	✓			✓		
	Curent Collector (Solid)		✓				✓				
	Cooling Channel (Fluid)	✓	✓								

Fig. 4.18 Governing Equations and zones where these are solved in Ansys Fluent

Fuel Cell Model Customization

The standard PEMFC module offers sub-models to capture various physical phenomena involved. These sub-models are based on correlations obtained from literature. However, in many scenarios, users prefer using their own sub-models developed from in-house testing. The PEMFC model in Ansys Fluent is offered in a user-defined function format through which the user can easily modify the pre-created pemfc_user.c file and include custom sub-models. For example, a leakage current value can be specified in the PEMFC model panel. However, if a user wants to specify custom correlation, that can be done using a pre-defined sub-routine shown in image below (Fig. 4.19).

Parameter Calibration

PEMFC model consists of many parameters which vary depending on the design of the individual components. Often, it may not be possible to derive values of all parameters from first principles or experimental data. To improve the model accuracy in such cases, the parameters must be calibrated to match the test data. Calibrating the

```
real Leakage_Current (cell_t c, Thread *t)
{
  real leak_i = leakage_current; /* unit:  A/m3 */

  /* The default constant value is "leakage_current",
   * which is taken from the user interface.
   * The default value may be overwritten here.
   */

  return leak_i;
}
```

Fig. 4.19 Customization of PEMFC model in ansys fluent

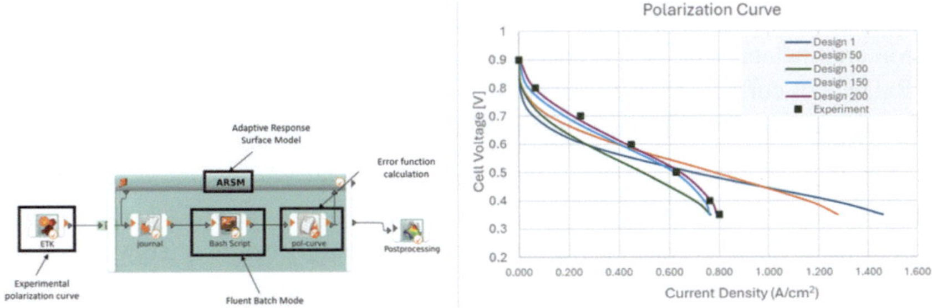

Fig. 4.20 Automated parameter calibration setup in OptiSLang (Left). Polarization Cuve changes with optimization iterations (Right)

parameters manually depends on user expertise and can be a laborious process. Ansys OptiSLang can be used to automate the calibration process. Experimental polarization curve is fed to OptiSLang. The optimization algorithm in OptiSLang minimizes the difference between experimental polarization curve and Fluent polarization curve by calibrating pre-selected parameters. The automated calibration process eliminates the need for a trial and error to obtain a set of parameters that match the experimental data (Fig. 4.20).

4.3.3   Software COMSOL Multiphysics

Scientists and engineers often use modeling and simulation in the exploratory stages of fuel cell research projects to understand the involved processes and phenomena.

The learning stage is often followed by prediction and validation. Scientists and engineers make predictions using the simulations and then try to validate these predictions with experiments. Validated models can be used to design new experiments and test new designs and operating conditions.

In this way, simulation becomes part of the innovation process, where significant improvements can be made in a design or a process. In the mature stages of research and development, simulation is used to optimize designs and operating conditions. These optimizations typically signify small incremental steps compared to changes made in the innovative stage, but over time these steps can yield major improvements.

The Fuel Cell & Electrolyzer Module contains a number of examples in its library of applications that show how to set up models and run simulations that are relevant in all stages in the research and development process.

Model of a PEMFC Unit

Let us look at one of these examples: a model of a PEMFC equipped with straight channels and operating under cocurrent flow of air and hydrogen. The model accounts for the:

- Multiphase flow in the gas channels as well as the porous electrodes
- Material balances in the gas channels, electrodes, and membrane
- Balance of current in the electrode (electronic conductor), pore electrolyte (ionic conductor), and membrane (ionic conductor)

The electrode kinetic expressions at both electrodes are functions of local gas composition and local overpotential. All of the involved equations — for the fluid flow, material balances, and balance of current — are fully coupled. They are solved for stationary and time-dependent studies. Figure 4.21 shows the model tree [80], which reflects the model definition, as well as the settings for the selected gas diffusion electrode and its corresponding oxygen-reduction reaction [81].

Figure 4.22 shows the model geometry. It consists of the gas channels, the electrodes, and the membrane. The metal plates make electronic contact with the electrodes, and the gas channels are formed by grooves in the metal plates. In Fig. 4.22, we see the gas channels in 3D and a 2D cross section of the cell perpendicular to the main flow direction in the gas channels. The inlets of air and hydrogen are placed on the same side of the fuel cell; i.e., the two gas streams are cocurrent.

In Fig. 4.23, we can see the oxygen mole fraction in the gas channels and cathode GDE at an average current density of 10 kA/m^2 and a cell voltage of about 0.59 V. We can see that there is a certain depletion of oxygen along the channels, from a mole fraction of 0.16 down to just below 0.1. The largest depletion occurs in the CL of the GDE. At this position close to the air outlet, the oxygen mole fraction decreases to a value of 0.05.

Figure 4.24 shows the corresponding hydrogen mole fraction plotted as the deviation from the inlet mole fraction of 0.69. The geometry is slightly rotated around a horizontal axis compared to Fig. 4.23 in order to get a view of the hydrogen channels. Figure 4.24 shows that there is only a very small depletion of hydrogen along the channels, so the cell is running with a high hydrogen excess.

In Fig. 4.25, we can see the resulting current density in the membrane. The figure shows that the current density is relatively uniform along the channels, with the highest value at the oxygen and hydrogen inlets. However, at the positions where the metal plate is in contact with the GDE, the current density is slightly lower. It is lowest at the position of this contact zone close to the outlets. The metal plate contact clearly obstructs the transport of oxygen. (The hydrogen composition from Fig. 4.24 shows that this has a very small impact.) If we made the channels wider and the contact zones between the GDE and the metal smaller, then this obstruction would be smaller. However, we would eventually reach a situation where the conduction of current becomes limiting to the sites farthest away from the contact zone, yielding lower current density in the middle of the channels. So, determining the optimal width and length of the channels depends on the material, the thickness of the electrodes, and the operating conditions.

Model of a PEMFC Stack

The Fuel Cell & Electrolyzer Module can also be used to model fuel cell stacks, in a fashion similar to the single cell model just demonstrated. In the following example, shown in Figs. 4.26 and 4.27, the flow channel configuration for a PEMFC stack is

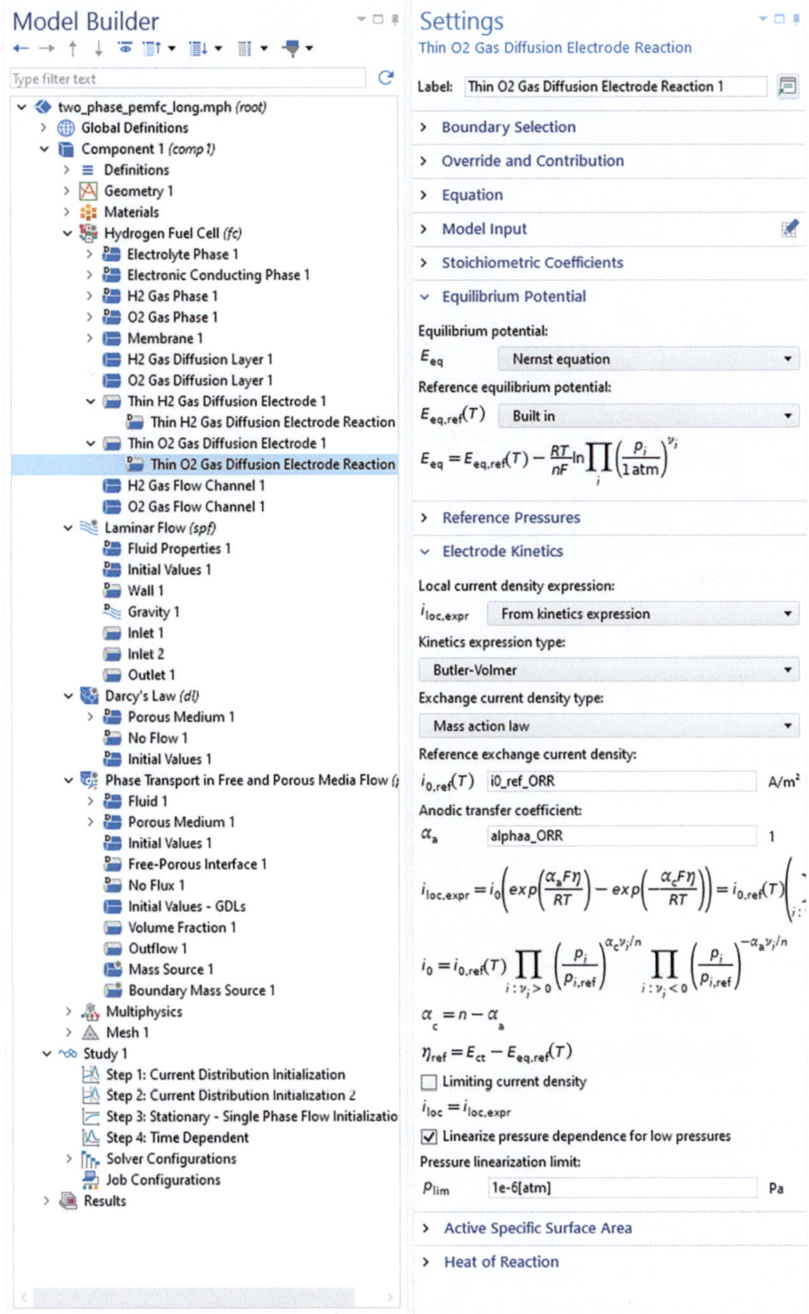

Fig. 4.21 The model tree in the Model Builder. The selected node in the tree defines the oxygen-reduction reaction occurring in the CL in the cathode GDE [80]

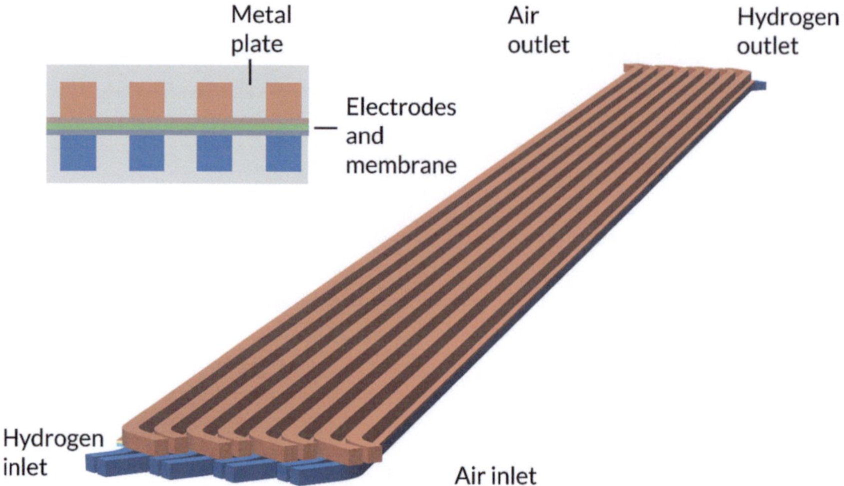

Fig. 4.22 A 3D drawing of the model geometry and a schematic cross section of the cell. The cross-section plane is perpendicular to the main direction of the flow in the gas channels. The channels are 20 cm in length

Fig. 4.23 Oxygen molar fraction in the gas channels and in the GDE cathode

shown for the hydrogen and air channels, respectively. The cell is also equipped with cooling channels, where water flows.

The oxygen supply is more limiting than the hydrogen supply in this case as well. In Fig. 4.28, we can see the mole fraction of oxygen. The current density distribution is largely governed by the oxygen supply (not shown). The oxygen mole fraction only varies to a small extent in each cell along the height of the stack.

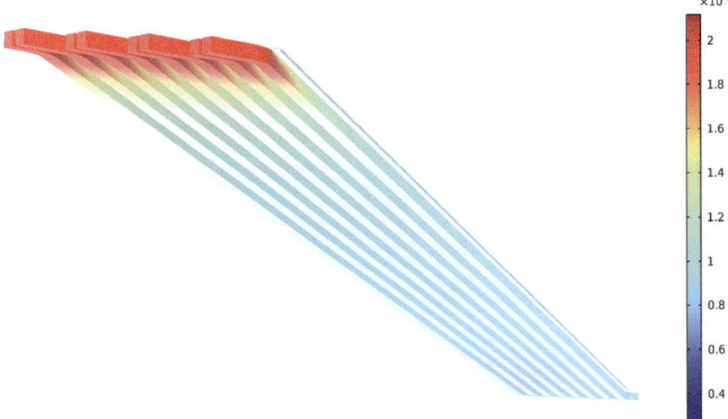

Fig. 4.24 Hydrogen mole fraction in the gas channels and in the anode GDE

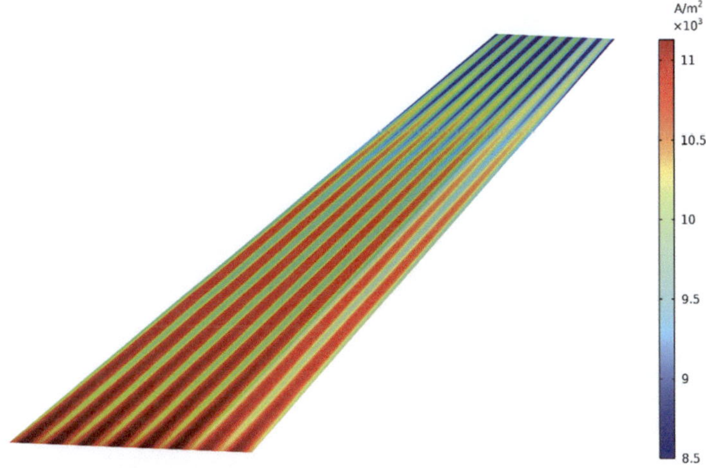

Fig. 4.25 Current density in the membrane at a cell voltage of approximately 0.59 V

The temperature field is shown in Fig. 4.29. We see that there is a substantial temperature variation in the stack. The highest temperature is in the cells farthest away from the cooling plate and close to the cooling water outlet. Within these cells, the highest temperature is found in the membrane. This is expected since Joule heating is a significant heat source and the membranes have poor electric and thermal conductivity compared with the other fuel cell components. This design is probably not optimal and would need to be improved to obtain a more uniform temperature.

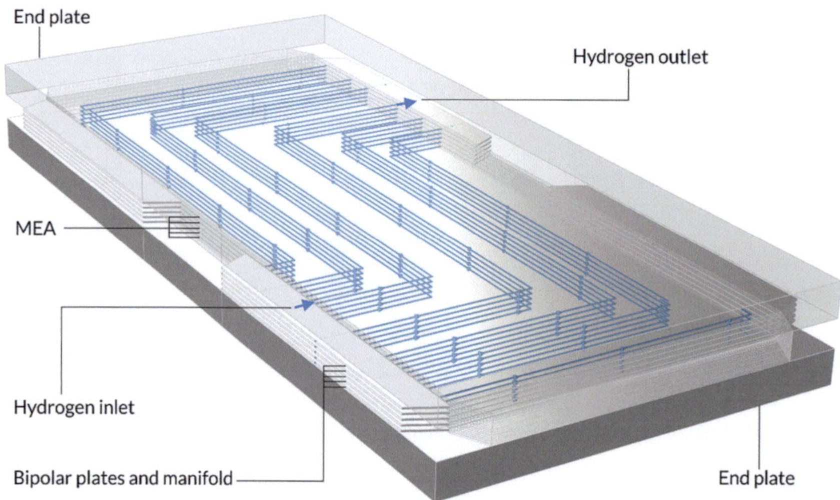

Fig. 4.26 Pattern for the hydrogen channels in a PEMFC stack. The stack fuel cell area is 24 cm in length and 10 cm in width

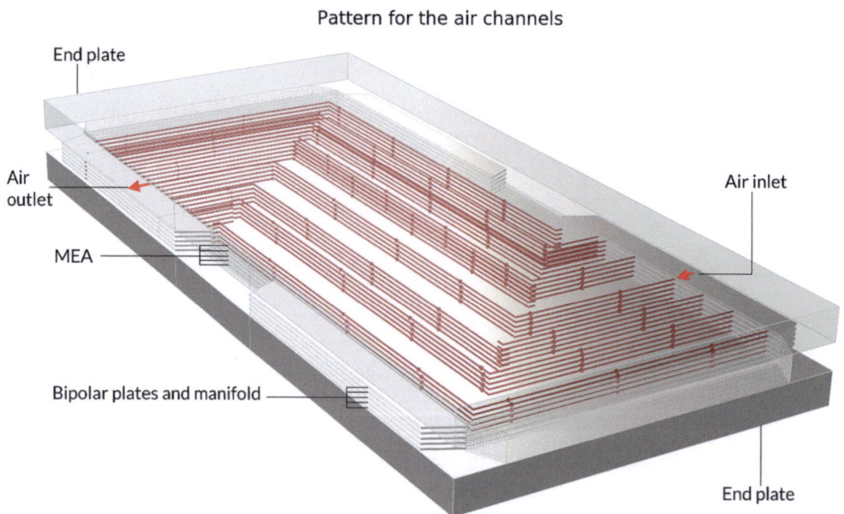

Fig. 4.27 Pattern for the oxygen channels in a PEMFC stack

Electrochemical Impedance Spectroscopy Study of a PEMFC

In fundamental studies, researchers and scientists may want to examine the phenomena in the GDE that limits the performance of the fuel cell. In such cases,

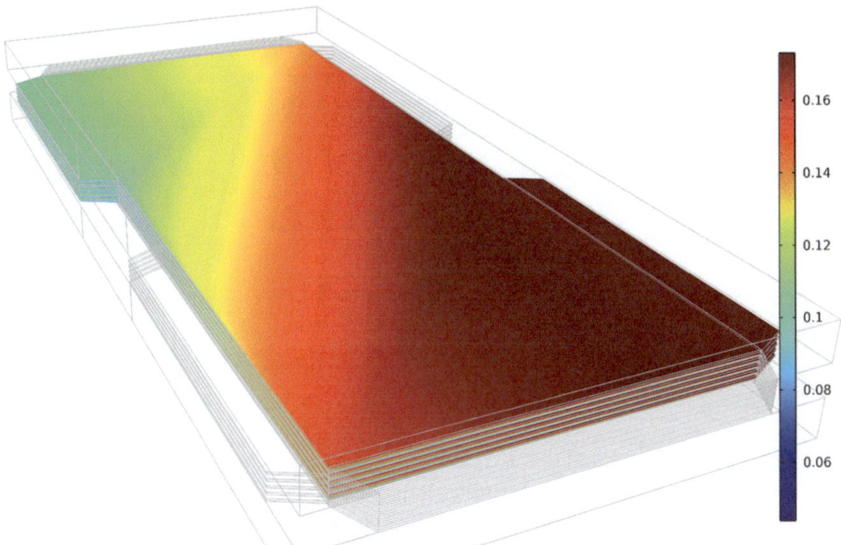

Fig. 4.28 Oxygen mole fraction in a PEMFC stack

Fig. 4.29 Temperature distribution in one cell in a fuel cell stack

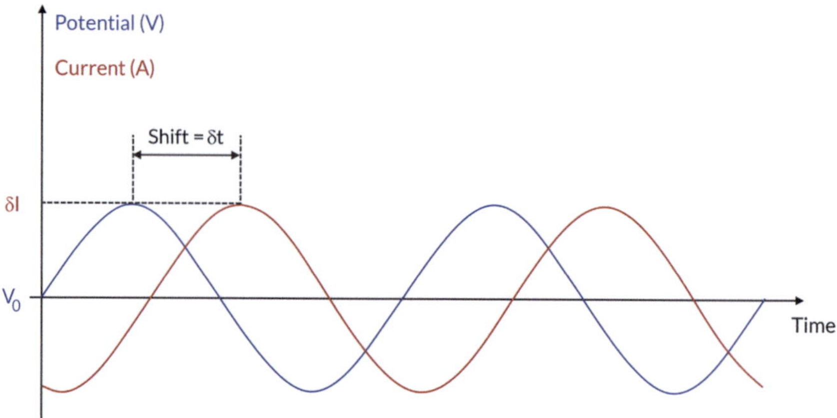

Fig. 4.30 Schematic principle of EIS. A small sinusoidal potential perturbation is applied on top of the cell's operating potential. The corresponding response in current is measured, and an impedance is computed. Note that in both cases, the sinusoidal potential and current are small parts of the total steady potential and current

EIS is often used. The idea is to apply a perturbation, for example, to the electric potential at the current collectors. This perturbation is typically sinusoidal in shape. The response of the current is then measured. The shift in time between the perturbation and the response at different frequencies tells us something about the time constant of the processes involved and how they may impact performance. The perturbation in the potential and the response of the current are used to compute a frequency-dependent impedance (Fig. 4.30).

In the Fuel Cell & Electrolyzer Module, the nonlinear part of the problem is solved first as a base level for the perturbation and response. This means that a small sinusoidal perturbation generates a small sinusoidal response in the current. If the perturbation in potential is small enough, then the response of the current is linear, allowing for a Laplace transform to reformulate the time-dependent problem to a linear problem in the frequency domain, with complex-valued field variables (transformed from real-valued variables). The resulting frequency-dependent impedance is thus also complex valued.

A large complex impedance at a given frequency represents a large loss and a long delay between perturbation and response. For example, at low frequencies, diffusion may be responsible for such a delay, as it is a slow process. A substantial increase in the impedance at low frequency may signify a large mass transport resistance. At high frequencies, effects of the electrode kinetics may yield a delay and a large contribution to the impedance.

The Fuel Cell & Electrolyzer Module features EIS as a built-in study type. This means that you can set up a stationary or time-dependent study and then add an EIS study to the same model with the same model settings. These types of models are often referred to as physics-based EIS models, in contrast to equivalent-circuit models.

The model in Fig. 4.31 treats a small portion of a fuel cell about 1 cm in length and 5 mm in width. The thickness of the whole geometry is 1 mm.

Figure 4.32 shows the flow field and the oxygen mole fraction in oxygen GDE. We can see that the oxygen mole fraction is slightly lower just below the position where the expanded mesh makes contact with the GDL. The cell voltage is 0.5 V, and the average current density is approximately 11.7 kA/m^2.

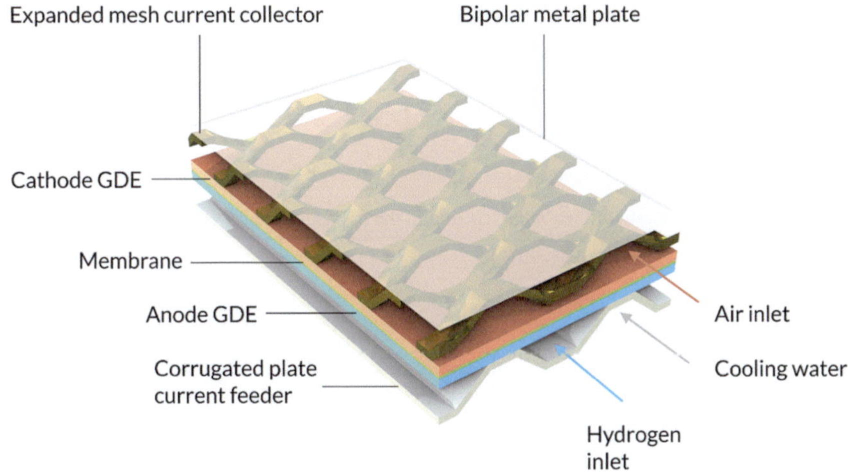

Fig. 4.31 Model geometry for an EIS model of a PEMFC unit

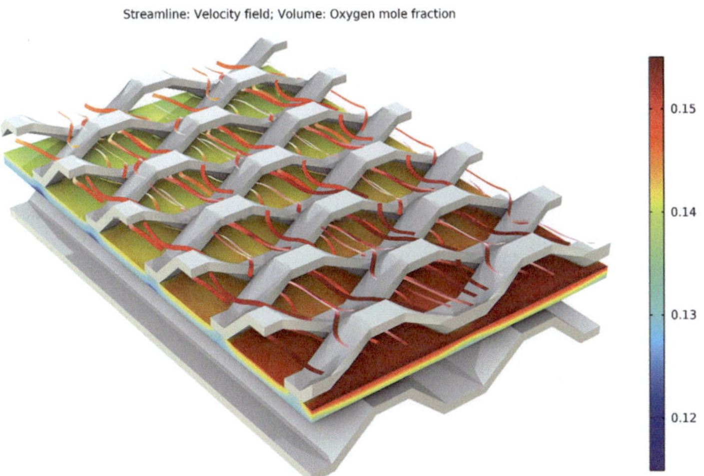

Fig. 4.32 Flow field and oxygen mole fraction in the oxygen GDE

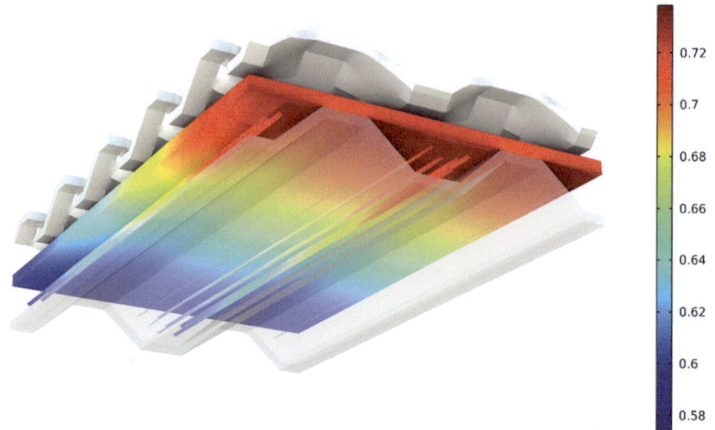

Fig. 4.33 Flow field and hydrogen mole fraction in the hydrogen channels and hydrogen GDE

If we look at the hydrogen anode (Fig. 4.33), we can see that the mole fraction decreases mostly due to the direction of the flow. There is a very small decrease at the position where the corrugated plate makes contact with the hydrogen GDE.

The resulting Nyquist plot for a perturbation of 5 mV and two different inlet velocities of air is shown in Fig. 4.34. In the two cases, the air velocity is 0.5 m/s and 1.0 m/s, respectively. As expected, the impedance is lower for the higher flow rate at low frequencies (to the right in the plot), as the higher flow rate decreases the mass transport resistance. At the right in Fig. 4.34, we can discern two small semicircles at low frequencies. The second semicircle at the lowest frequency (the second semicircle from the right) represents the transport of oxygen, while the first semicircle originates from a change in the current density distribution with frequency, due to three-dimensional effects. The largest semicircle (the first semicircle from the left) represents the kinetics of the oxygen-reduction reaction. The hydrogen reaction kinetics and the hydrogen-transport resistances are so small that they are hidden below the contributions from the oxygen reduction; the overvoltage from the oxygen reaction is 370 mV, while it is only about 5 mV for the hydrogen oxidation. However, we could also plot the impedance for the two electrodes separately, toward a reference. Then, we could also investigate the hydrogen transport and reaction in detail.

In a similar way, other operating conditions can be varied in the model in order to understand how the variations impact the performance, which is revealed in the Nyquist plots. For example, different materials may have different catalytic activity, which would typically be revealed at higher frequencies in the Nyquist plot, as in Fig. 4.35.

Figure 4.35 shows the Nyquist plot for two different values of the exchange current density (two different rate constants) for the oxygen-reduction reaction. In this case, since we are running the potential perturbation, the perturbation current

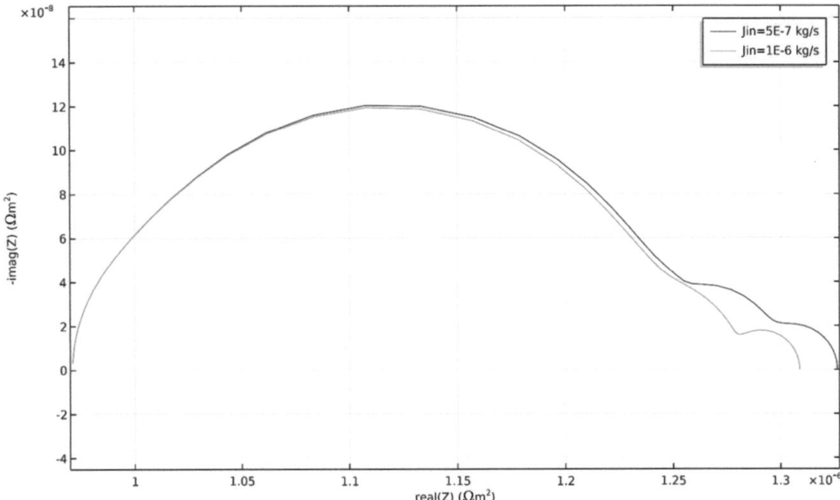

Fig. 4.34 Nyquist plot for two different air flow velocities

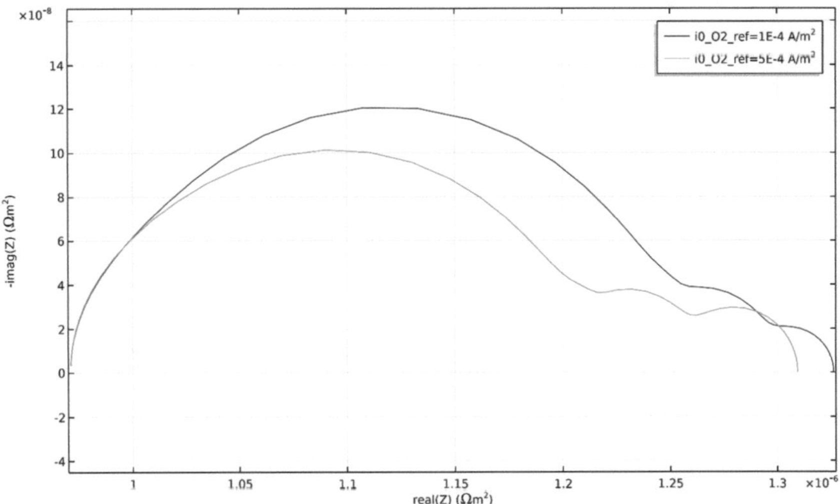

Fig. 4.35 Nyquist plot for two different exchange current densities for oxygen reduction

also increases slightly as the exchange current density increases. This leads to a smaller kinetic impedance (large semicircle) but a somewhat larger mass-transport impedance (the second smaller semicircles), which is clearly seen in the oxygen transport semicircle (second from the right).

Scientists and engineers can use EIS models to design experiments and extract properties of the electrodes and electrolyte in the fuel cell at different operating conditions, as demonstrated. They can also use EIS to evaluate the state of health of a fuel cell and determine if it is running under kinetic, mass transport, or ohmic limitations.

Chapter 5
Practical Considerations for PEMFC CFD Modelling

In this chapter to practical examples for applications of PEM fuel cell modelling will be discussed. The applications are within the industrial sector as larger active areas of about 140 cm^2 are modelled. In addition, practical tips are provided in Sect. 5.2 for a more direct usage for the readers, based on the experience of the authors in CFD modelling PEM fuel cells.

5.1 Applications in Industry and Research

CFD is widely used for research purposes as well as for product development in industry. The scope of the simulations carried out therefore may be different for each case, and obviously also the difference in the length-scale or dimensions of the cells being modelled.

An example of a 140 cm^2 cell is presented in this section, where the objective was to design a new bipolar plate and flow field for improved water management [82]. The bipolar plates are made of 0.1 mm thick 316L stainless steel, with a 20 nm thickness coating applied by PVD (Physical Vapour Deposition) to prevent the metal sheet from corroding. Both anode and cathode sides of the plates are glued together using reinforced epoxy papers (prepregs), with the coolant flow field in between both plates. The respective gas and cooling manifolds are included. The MEA being used is a Gore M820.15 3L-MEA where GDLs are assembled to the framed 3L-MEA. GDLs present a nominal thickness 245 mm and 78% porosity and include Micro Porous Layer (MPL). In the corresponding CFD simulation orthotropic properties (thermal conductivity and electric conductivity) were considered in the GDL. The flow field layout of the cell is crossflow where the cathode channels present a vertical parallel flow field design, whereas the anode channel is a horizontal single serpentine. The cooling water flows from top-right towards bottom-left.

A. Iranzo et al., *Computational Fluid Dynamics Modelling of PEM Fuel Cells*, SpringerBriefs in Energy, https://doi.org/10.1007/978-3-032-06631-2_5

The details on the CFD model are described in [82]. The in-plane liquid water saturation through CFD simulation results is depicted in Fig. 5.1 for different operating conditions. It can be observed that as with the increase in the intensity, the liquid water content increases reaching maximum values in the lower part of the cell (due to the fact that gases become saturated with moisture as they advance along the flow paths towards the outlet). In a side-view of the cell, a larger amount of liquid water is observed under the ribs than under the channels on both the anode and cathode sides (Fig. 5.1), which is a well-known feature caused by the fact that lands have the sufficient liquid water present and the coldest GDL locations because of the water cooling.

Optimized anode, cathode and coolant flow design for enhanced performance of PEMFC.

The flow field channels design in a PEMFC plays a crucial role in output fuel cell performance. Firstly, the BPP component is highly responsible to evenly distribute the reactant feed gasses to the catalyst surface for electrochemical reactions. The main advantage of uniform gas distribution is that it helps in preventing formation of hot spots and improve the durability and efficiency of fuel cell. Secondly, the flow field also required to remove the un-reacted gases from the reaction site. Moreover, the liquid water formed as by product must be also effectively transported away, or

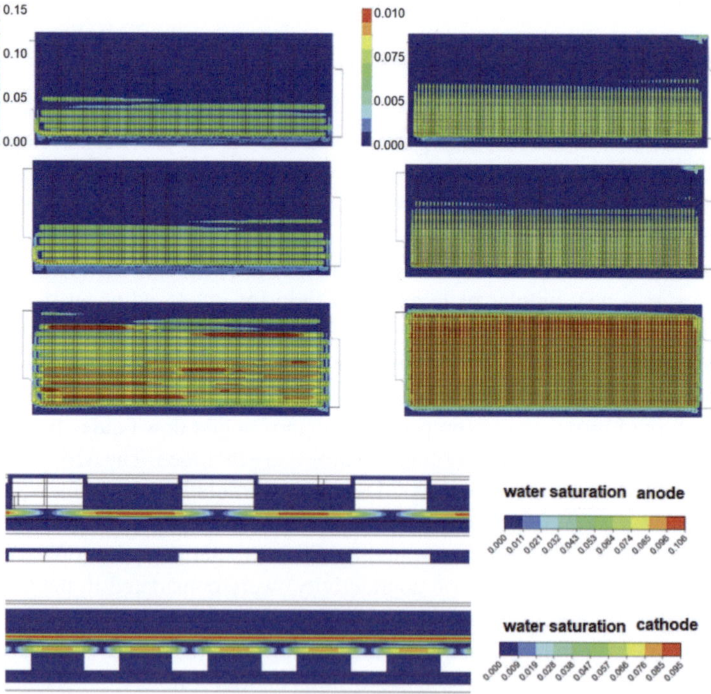

Fig. 5.1 Liquid water distribution in plan and through plane for anode and cathode GDL

flooding effect is observed resulting in reduced fuel cell performance. Thirdly, the flow field design must also be optimized to minimize the contact resistance between bipolar plate and GDL. The reaction is an electrochemical process, and the flow field must be designed to facilitate the transport of electrons from the reaction site to the external circuitry and mechanical support. Finally, the flow field must be designed to minimize the pressure drop and evenly distribution of pressure from inlet to outlet. A large pressure drop can cause to deploy more powerful external blowers to force the gas through the system increasing the overall cost in operation to the system as well as induce a parasitic load which reduces the overall efficiency of the system. These flow field features such as uniformly distribute the gases, transport and removal of product water, minimize contact resistance, while minimizing overall drag make it a challenging task to design an optimized flow field. Although, the efficiency and performance of PEMFCs can be enhanced by optimizing the flow field geometry where the bipolar plates are considered as important component of the fuel cell as it helps in distribution of gases from the inlet manifold and through gas channels to the electrode surface for electrochemical reaction, and removal of produced water in the gas channels through outlet manifold with a desired pressure drop value. Also, bipolar plates take a considerable share in overall fuel cell manufacturing cost, to mitigate cost and improve performance as the bipolar plates with optimum design parameters and corresponding pressure drop [83].

Following the best practices for ANSYS PEMFC modelling and simulation process, the model mesh is divided into two separate sections as shown in Fig. 5.2. First, the MEA section is meshed using the quadrilateral and hexahedral elements with 10-5-5-5-10 ratio and second, the bipolar plates of anode and cathode are meshed through the watertight geometry workflow, where the solid and fluid regions of the bipolar plates can be assigned. The mass flow type is assigned to anode and cathode inlets, velocity flow type is used for the coolant inlets for anode and cathode sides, whereas the outlets for the mass flow and velocity flow type are defined as pressure outlets. The mesh is in these components are filled using polyhedral elements with a reporting orthogonal quality (0.1) and the mesh statistics are shown in Table 5.1, merged creating a non-conformal interface between the bipolar plates and MEA [84].

Water uptake is crucial for PEM membrane, especially on the cathode side, without creating excess water flooding to promote reactants on to reaction zone for electro-chemical reaction and the inadequate membrane water content can sometimes lead to drying and affecting the proton conductivity reducing the overall fuel cell perfor-mance. To ensure the optimum membrane water content, the water formation in cathode flow field designs is analyzed as shown in Fig. 5.3. In this application, the homogeneity of water content distribution on the membrane surface has been consid-ered as one of the key selection criterions. From the contour plots shown in Fig. 5.3, the average membrane water content estimated at 0.61 V (a, d and g) for the multi-parallel serpentine (inlet and outlet on opposite sides), multi-parallel serpentine (inlet and outlet on same side), and multi-serpentine are 8.9, 9 and 8.5, respectively. It is well known that more heat can be observed under the channel section than the rib section, which will have an impact on temperature distribution and ultimately can have a significant effect on evaporation of water and adequate hydration of membrane

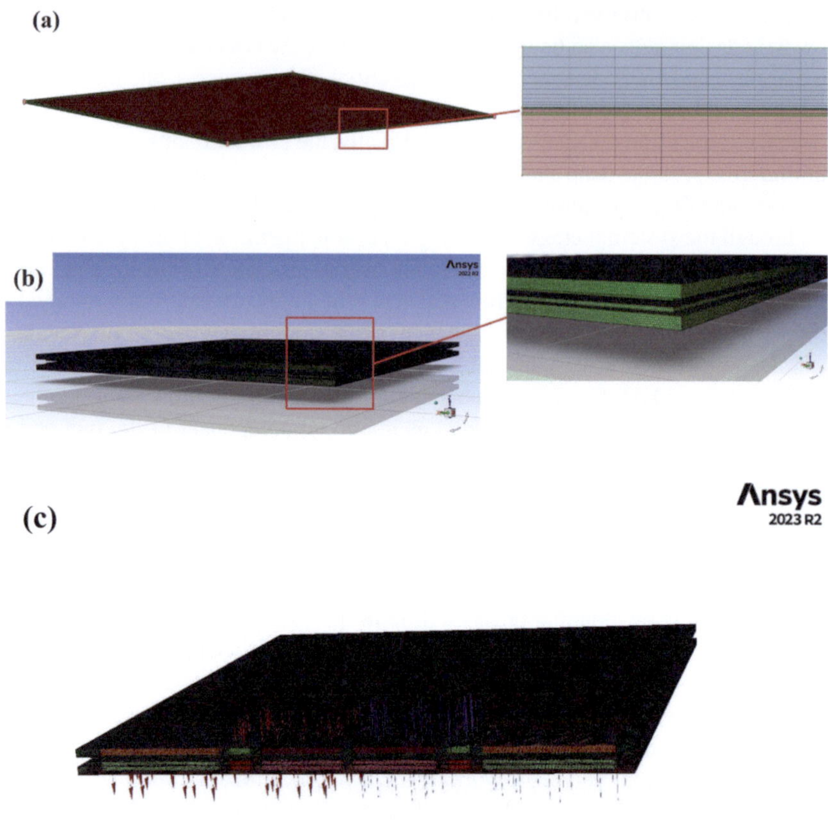

Fig. 5.2 Sectional view of **a** Membrane electrode assembly and **b** BPP mesh **c** Appended case file (BPP and MEA) for simulation

Table 5.1 Mesh statistics for the three cells

	Cell 1	Cell2	Cell 3
MEA divisions	10-5-5-5-10	10-5-5-5-10	10-5-5-5-10
BPP elements	9,316,863	8,532,445	8,629,576
MEA elements	4,393,830	4,419,450	4,422,600

for proton conduction. The cell generates more water at high current densities, here the average membrane water contents for Cell 1 to 3 (b, c and h) at 0.41 V are 9.74, 10.38 and 10.36, respectively. Cold patches are observed in Cell 3 (h, i) at higher current densities resulting in poor distribution and dehydration of membrane. Among the combination of flow field designs, Cell 2 (d, e and f) has the uniform distribution of water content on the active area, which showed an improved output performance (0.97 W.cm^{-2}) and profound resistance to mass transfer losses [83].

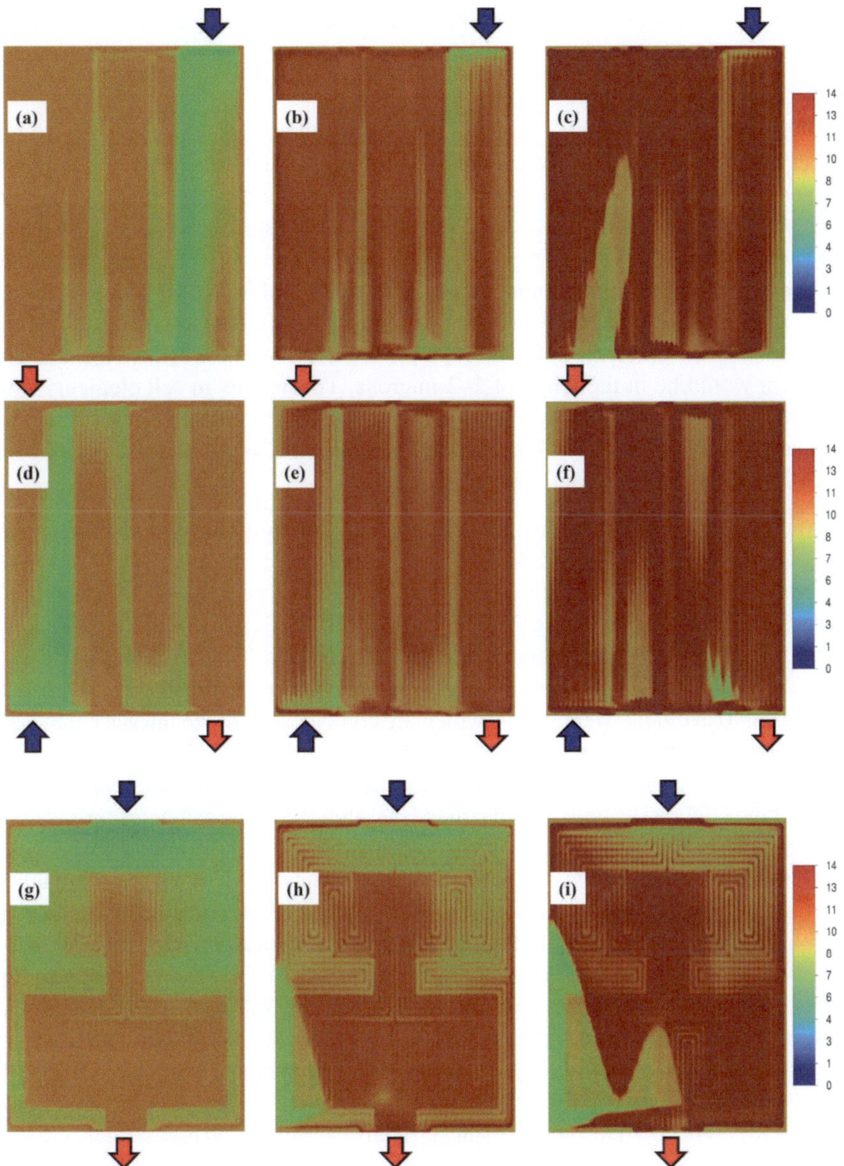

Fig. 5.3 Membrane water content distribution for different flow designs [83]

5.2 Considerations and Practical Hints for Modelling and Simulation of PEM Fuel Cells

The following considerations and practical hints are based on the experience of the authors in PEMFC CFD modelling over the last 15 years, and cover meshing strategies, model set-up and simulation strategies, convergence issues, and other practical items.

- Meshing considerations: hexahedral mesh is absolutely recommended due to the very high aspect ratio of the geometries involved in PEM fuel cells, in particular membrane and catalyst layers (when modelled as 3D media, which is required depending on the software used). Metal bipolar plates produced from thin metal sheets (100 micros or less) presents a similar situation. A typical 100 cm^2 active area cell with Nafion membrane of about 20 microns thickness is thus presenting an aspect ratio of 1:10,000. In addition, at least 10 mesh elements shall be resolving the membrane gradients in the through-plane direction of the cell, so the element height would be in the range of 1–2 microns. This results in cell elements with high aspect ratios. While most CFD software does handle such elements properly, double precission solver will be required in some case for mesh aspect ratios over 1000 depending on the software being used.

Another meshing strategy which is suitable and required for complex flow field layouts and different anode and cathode flow field designs is the use of non-conformal meshes or mesh-interfaces, so that different meshes can be generated for each cell side and then connected into the CFD Pre-processor. This is suitable and significantly simplifying the mesh generation for complex cell designs. The decision on where to place the mesh interface must be obviously made before generating the mesh, and it is convenient to avoid mesh interfaces which are coincident with the interface between cell components (membrane-electrode interface, electrode-GDL interface, GDL-rid/gas channel interface). Placing the interface in the mid-place of the membrane or mid-plane of the GDL is strongly recommended, as depicted in the examples provided in Fig. 5.4.

Mesh independence analysis is absolutely recommended to ensure the quality and trust of PEMFC CFD modelling, as in any other CFD application. The analysis shall be carried out according to the regular CFD Best Practice Guidelines [4], and/or following well-stablished methodologies such as the Grid Convergence Index (GCI) [86]. This is not always present in publications in the field, although it is certainly required for ensuring the quality of the results. Based on the experience of the authors, such methodologies provide in general the fact that at least over 100,000 cell elements must be used for each cm^2 of active area of the cell (for single cell models) [87]. The mesh resolution in the through-plane direction is absolutely critical for the accuracy of the results, when high gradients of key variables such as water saturation are present. At least 10 elements should be present for each cell component in the through-plane direction (membrane, electrodes, GDLs, flow field channels).

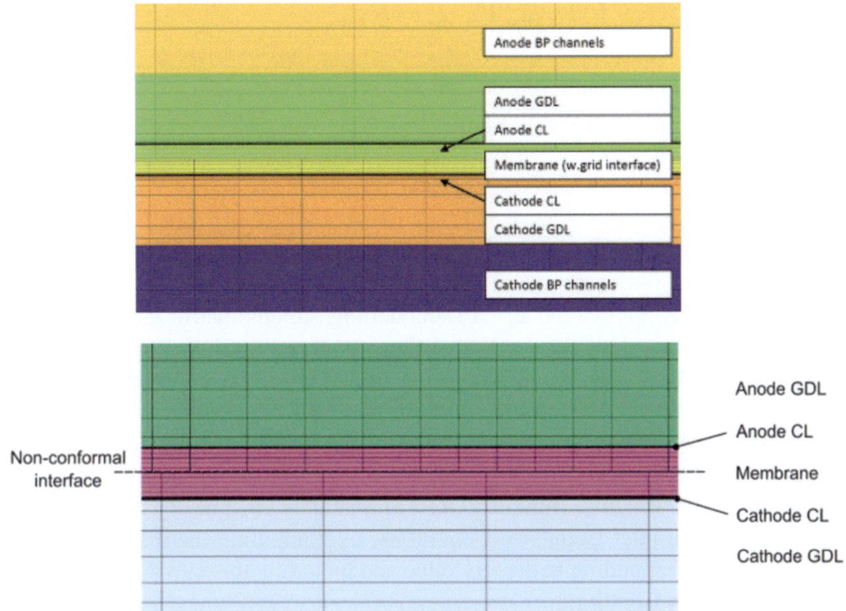

Fig. 5.4 Cross section (through-plane) views of hexahedral meshes featuring non conformal mesh interfaces at the membrane

For the Bipolar Plate, where heat conduction and charge transport is resolved, the number of elements in the through-plane direction can be reduced to 4–5.

Finally, it is highly recommended that the mesh transition between cell components must be smooth, i.e. avoiding cell element heights of 1 micron in the electrode next to a cell element with height of 20 microns in the GDL in contact with the electrode. A smooth transition in cell sizes will greatly contribute to the robustness of the solver and to the accurate resolution of the different variable gradients.

- Model set-up: fuel cell stack systems operate at a certain stoichiometric flow of reactants. Stoichiometry refers to the flow of reactants (kg/s of hydrogen or oxygen) with respect to the kg/s strictly required for a certain current intensity. The relationship between current and reactant flow is given by the Faraday's Law. For the anode flow (hydrogen) the Faraday's Law reads:

$$m_{H_2} = \frac{M_{H2}}{2F} I_{ref} A_{cell}$$

- where, A_{cell} is the cell active area; F is Faraday constant (96 485.3321 s A/mol); and m_H2 is the mass flux at the anode inlet, which can be calculated from the total mass flow rate and the hydrogen mass fraction at the inlet:

$$m_{H2}^{in} = m_{H2}^{anode} * y_{H2}$$

The equivalent expression for the cathode oxygen flow reads:

$$\xi_{O2} = \frac{m_{O2}^{cathode} * y_{O2}}{\frac{M_{O2}}{4F} I_{ref} A_{cell}}$$

Typically fuel cells operates at anode stoichiometry 1.2–1.5 and cathode stoichiometry 1.5–2.5 [88].

In the practical application in CFD modelling stoichiometric factors must be accounted for when defining the inlet Boundary Conditions for both anode and cathode flows. Based on the current being drawn from the cell and imposed as Boundary Condition (galvanostatic condition) the stoichiometric mass flow rate of hydrogen and oxygen can be computed from the Faraday's Law, and then multiplied by the stoichiometric factors to determine the actual flow rates of reactants.

In addition, the Relative Humidity (RH) of reactants must be considered as well when defining the inlet Boundary Conditions, by computing the mass fraction of water vapour corresponding to the RH value and including this into the definition of the inlet Boundary Condition.

- Simulation strategies: typically, the polarization curve or I-V curve is computed for a cell or a stack. For this, different points of the curve are calculated by updating the corresponding Boundary Conditions, either potentiostatic or galvanostatic, and adjusting the flow rates of reactants as appropriate. It is highly recommended to start the simulations from the point with the lowest current density, and after convergence use the results file as initial conditions for the simulations of the next points, increasing the current density. This is because liquid water generation will be more relevant at high current densities, and liquid water transport is typically one of the items presenting convergence issues, so at high current densities it will help the solver to have as initial conditions the results from a slightly lower current density. Therefore, when using potentiostatic boundary conditions (fixed voltage), the simulations can start with a high voltage and gradually decrease it. On the other hand, for galvanostatic boundary conditions (fixed current density), simulations can start with a low current density and gradually increase it.
- Convergence: In case of convergence difficulties even with the low current density cases, it is useful to simplify the model complexity by turning-off some of the equations, such as the energy sources (Joule heating and reaction heating) for some iterations until the flow field and the species and electric potential distributions are well developed. It may be also necessary to turn-off the multiphase flow model to obtain first a well-developed solution for flow, species and potentials, and then activate back the multiphase flow model.

Obviously, the usual convergence strategies for each software being used apply also for Fuel Cell simulations (reducing timescales, reducing Under-Relaxation Factors, using first-order/upwind discretization schemes for aiding convergence and then swith to second-order schemes for achieving accuracy). The previous general hints can be particularly useful for equations with convergence difficulties, which are typically the ones related to water transport.

In addition to determine convergence by the regular approaches (residuals, imbalances, and steady-state values of variables with respect to the iteration number) it is highly recommended to evaluate whether the overall charge transport (current intensity) is corresponding to the reactants consumption (hydrogen and oxygen) and to the water being generated in the cell. For this, the Faraday's Law for hydrogen and oxygen can be used. Current being drawn from the cell must correspond to a certain consumption of hydrogen and oxygen, according to Faraday's Law) and this must be verified by monitoring the inlet and outlet flow of reactants (so that the difference is the consumed flow of reactants that must correspond to the value provided by the Faraday's Law for each current). In the same manner, the water being generated in the cell (electrochemically) which can again be determined from the difference between the outlet and inlet flows (considering anode and cathode) must correspond to the overall cell reaction:

$$H2 + {}^1\!/_2\, O2 \rightarrow H2O$$

Once this is fulfilled there is a guarantee that the simulation is properly converged.

Chapter 6
Conclusions

This book briefly describes the different fuel cell technologies based on various operating parameters such as temperature range, electrolyte, power density, and their respective applications from small to large scale. The fundamentals of Computational Fluid Dynamics numerical simulation are presented as well as particularly detailed for PEMFC, which can be applicable to other types of fuel cells. The wider use of PEMFCs in industry and academic research made such technology suitable for a cost-effective CFD-aided design optimization. It is worth noting that the coupled fluid flow to complex interactions between heat and mass transport as well as electrochemical reactions make overall fuel cell process difficult to predict. CFD modelling is a strongly suitable method to predict such complex interplay between transport and reactions, as well as to assess design decisions and to optimize operating conditions. 1D and 2D based models are cost-effective and do not need much computational resources, but can be applied to understand only a particular component mechanism making 3D models a necessity to the multi-interaction phenomena in fuel cell components.

The accuracy of CFD simulation models is still dependent on experimental results for fitting some key parameters such as interfacial contact, reference exchange current density and electrode properties, making it a challenging for validation.

Unique features of major commercial software for PEM fuel cell modelling and simulation such as AVL FIRE, ANSYS, or COMSOL Multiphysics are highlighted in this book, with a comprehensive and useful source of information for all practitioners. The final consideration and hints provided in the book are also expected to become a useful source of information for CFD users facing PEMFC modelling.

© The Author(s), under exclusive license to Springer Nature Switzerland AG 2025   69
A. Iranzo et al., *Computational Fluid Dynamics Modelling of PEM Fuel Cells*,
SpringerBriefs in Energy, https://doi.org/10.1007/978-3-032-06631-2_6

References

1. Sharaf OZ, Orhan MF (2014) An overview of fuel cell technology: fundamentals and applications. Renew Sustain Energy Rev 32:810–53. https://doi.org/10.1016/j.rser.2014.01.012
2. Ahmad S, Nawaz T, Ullah A, Ahmed M, Khan MO, Saher S, et al (2020) Thermal optimization of manganese dioxide nanorods with enhanced ORR activity for alkaline membrane fuel cell. Electrochem Sci Adv e2000032. https://doi.org/10.1002/elsa.202000032
3. Alkaline Membrane Fuel Cell Workshop|Department of Energy (n.d)
4. Barton SC, Cells EF (2009) Chapter 8 Enzyme catalysis in biological fuel cells 1 EMERGING APPLICATIONS FOR. Enzyme 5:1–19
5. Gouérec P, Poletto L, Denizot J, Sanchez-Cortezon E, Miners JH (2004) The evolution of the performance of alkaline fuel cells with circulating electrolyte. J Power Sources 129:193–204. https://doi.org/10.1016/j.jpowsour.2003.11.032
6. Litster S, McLean G (2004) PEM fuel cell electrodes. J Power Sources 130:61–76. https://doi.org/10.1016/j.jpowsour.2003.12.055
7. Shi X, Ahmad S, Pérez-Salcedo K, Escobar B, Zheng H, Kannan AM (2019) Maximization of quadruple phase boundary for alkaline membrane fuel cell using non-stoichiometric α-MnO2 as cathode catalyst. Int J Hydrogen Energy 44:1166–1173. https://doi.org/10.1016/J.IJHYDENE.2018.11.042
8. Reverdiau G, Le Duigou A, Alleau T, Aribart T, Dugast C, Priem T (2021) Will there be enough platinum for a large deployment of fuel cell electric vehicles? Int J Hydrogen Energy 46:39195–207. https://doi.org/10.1016/j.ijhydene.2021.09.149
9. Rodrigues A, Amphlett JC, Mann RF, Peppley BA, Roberge PR (1997) Carbon monoxide poisoning of proton-exchange membrane fuel cells. In: IECEC-97 proceedings of the thirty-second intersociety energy conversion engineering conference (Cat No97CH6203), vol 2, pp 768–73. https://doi.org/10.1109/IECEC.1997.660236
10. Molochas C, Tsiakaras P (2021) Carbon monoxide tolerant pt-based electrocatalysts for h2-pemfc applications: current progress and challenges. Catalysts 11. https://doi.org/10.3390/catal11091127
11. Yang Z, Wang B, Jiao K (2020) Life cycle assessment of fuel cell, electric and internal combustion engine vehicles under different fuel scenarios and driving mileages in China. Energy 198:117365. https://doi.org/10.1016/j.energy.2020.117365
12. Eapen DE, Suseendiran SR, Rengaswamy R (2016) 2—Phosphoric acid fuel cells. In: Barbir F, Basile A, Veziroğlu TNBT-C of HE (eds) Woodhead Publishing Series in Energy, Oxford: Woodhead Publishing, p 57–70. https://doi.org/10.1016/B978-1-78242-363-8.00002-5
13. Neergat M, Shukla AK (2001) A high-performance phosphoric acid fuel cell. J Power Sources 102:317–21. https://doi.org/10.1016/S0378-7753(01)00766-2

A. Iranzo et al., *Computational Fluid Dynamics Modelling of PEM Fuel Cells*, SpringerBriefs in Energy, https://doi.org/10.1007/978-3-032-06631-2

14. Mamlouk M, Scott K (2010) The effect of electrode parameters on performance of a phosphoric acid-doped PBI membrane fuel cell. Int J Hydrogen Energy 35:784–93. https://doi.org/10.1016/j.ijhydene.2009.11.027

15. Okumura molecular sciences and chemical engineering MBT-RM in C. FUEL CELLS-PHOSPHORIC ACID FUEL CELLS|Systems⋆, Elsevier (2013). https://doi.org/10.1016/B978-0-12-409547-2.01235-X

16. Martín AJ, Hornés A, Martínez-Arias A, Daza L (2013) Recent advances in fuel cells for transport and stationary applications. Renew Hydrog Technol Prod Purification, Storage, Appl Saf, Elsevier B.V 361–80. https://doi.org/10.1016/B978-0-444-56352-1.00015-5

17. Ghouse M (1999) Effect of the long term testing of the phosphoric acid fuel cell stacks on platinum catalyst crystallite size. J New Mater Electrochem Syst 2

18. Ma Z, Venkataraman R, Farooque M (2009) Fuel cells—molten carbonate fuel cells|modeling. In: Garche JBT-E of EPS (ed) Amsterdam: Elsevier, p 519–32. https://doi.org/10.1016/B978-044452745-5.00272-0

19. Behling NH (2013) Chapter 2—fuel cells and the challenges ahead. In: Behling NHBT-FC (ed). Elsevier, p 7–36. https://doi.org/10.1016/B978-0-444-56325-5.00002-8

20. Cassir M, Ringuedé A, Lair V (2013) 17—molten carbonates from fuel cells to new energy devices. In: Lantelme F, Groult HBT-MSC (eds) Oxford: Elsevier, p 355–71. https://doi.org/10.1016/B978-0-12-398538-5.00017-2

21. Uzunoglu M, Alam MS (2018) 33—fuel-cell systems for transportations. In: Rashid MHBT-PEH, Fourth E, editor, Butterworth-Heinemann, p 1091–112. https://doi.org/10.1016/B978-0-12-811407-0.00037-4

22. Di Giulio N, Bosio B, Han J, McPhail SJ (2014) Experimental analysis of SO2 effects on molten carbonate fuel cells. Int J Hydrogen Energy 39:12300–8. https://doi.org/10.1016/j.ijhydene.2014.04.120

23. Biedenkopf P, Spiegel M, Grabke HJ (1997) High temperature corrosion of low and high alloy steels under molten carbonate fuel cell conditions. Mater Corros 48:477–88. https://doi.org/10.1002/maco.19970480803

24. Jörissen L, Gogel V (2009) Fuel cells – direct alcohol fuel cells|direct methanol: overview. In: Garche JBT-E of EPS. Amsterdam: Elsevier, p 370–80. https://doi.org/10.1016/B978-044452745-5.00241-0

25. Goor M, Menkin S, Peled E (2019) High power direct methanol fuel cell for mobility and portable applications. Int J Hydrogen Energy 44:3138–43. https://doi.org/10.1016/j.ijhydene.2018.12.019

26. Dohle H, Mergel J, Stolten D (2002) Heat and power management of a direct-methanol-fuel-cell (DMFC) system. J Power Sources 111:268–82. https://doi.org/10.1016/S0378-7753(02)00339-7

27. Stauffer DB, Hirschenhofer JH, Klett MG, Engleman RR (1998) Fuel cell handbook, 4th edn. Pittsburgh, PA, and Morgantown, WV (United States). https://doi.org/10.2172/14997

28. Tewary A, Upadhyay C (2022) Fuel cell technology: the future ahead BT—environmental concerns and remediation. In: Ashish DK, de Brito J. Cham: Springer International Publishing, p 249–79

29. Song SQ, Zhou WJ, Li WZ, Sun G, Xin Q, Kontou S et al (2004) Direct methanol fuel cells : methanol crossover and its influence on single DMFC performance. Ionics (Kiel) 10:458–462. https://doi.org/10.1007/BF02378008

30. Zhou J, Cao J, Zhang Y, Liu J, Chen J, Li M et al (2021) Overcoming undesired fuel crossover: goals of methanol-resistant modification of polymer electrolyte membranes. Renew Sustain Energy Rev 138:110660. https://doi.org/10.1016/J.RSER.2020.110660

31. Sharifzadeh M, Shah N (2020) Chapter 9 - Fuel variability and flexible operation of solid oxide fuel cell systems. In: Sharifzadeh MBT-D and O of SOFC. Woodhead Publishing Series in Energy, Academic Press, p 277–95. https://doi.org/10.1016/B978-0-12-815253-9.00009-4

32. Irshad M, Siraj K, Raza R, Ali A, Tiwari P, Zhu B, et al (2016) A brief description of high temperature solid oxide fuel cell's operation, materials, design, fabrication technologies and performance. Appl Sci 6. https://doi.org/10.3390/app6030075

33. Boldrin P, Ruiz-Trejo E, Mermelstein J, Bermúdez Menéndez JM, Ramírez Reina T, Brandon NP (2016) Strategies for carbon and sulfur tolerant solid oxide fuel cell materials, incorporating lessons from heterogeneous catalysis. Chem Rev 116:13633–84. https://doi.org/10.1021/acs.chemrev.6b00284

34. Vinchhi P, Khandla M, Chaudhary K, Pati R (2023) Recent advances on electrolyte materials for SOFC: a review. Inorg Chem Commun 152:110724. https://doi.org/10.1016/j.inoche.2023.110724

35. Sargent and Lundy (2020) Capital cost and performance characteristic estimates for utility scale electric power generating technologies. Washington, DC

36. Ramsden T (2013) An evaluation of the total cost of ownership of fuel cell powered material handling equipment

37. Kampker A, Ayvaz P, Schön C, Karstedt J, Förstmann R, Welker F (2020) Challenges towards large-scale fuel cell production: results of an expert assessment study. Int J Hydrogen Energy 45:29288–96. https://doi.org/10.1016/j.ijhydene.2020.07.180

38. Jiao K, Xuan J, Du Q et al (2021) Designing the next generation of proton-exchange membrane fuel cells. Nature 595:361–369. https://doi.org/10.1038/s41586-021-03482-7

39. Qasem NAA, Abdulrahman GAQ (2024) A recent comprehensive review of fuel cells: history, types, and applications. Int J Energy Res 2024(1):7271748. https://doi.org/10.1155/2024/7271748. and references cited therein

40. Date AW (2005) Introduction to computational fluid dynamics. Cambridge University Press

41. Anderson JD, Wendt J (1995) Computational fluid dynamics, vol 206, p 332. New York: McGraw-Hill

42. Sharma A (2021) Introduction to computational fluid dynamics: development, application and analysis. Springer Nature

43. Iranzo A (2019) CFD applications in energy engineering research and simulation: an introduction to published reviews. Processes 7(12):883

44. Maher AR, Sadiq AB (2008) CFD models for analysis and design of PEM fuel cells

45. Kone JP, Zhang X, Yan Y, Hu G, Ahmadi G (2017) Three-dimensional multiphase flow computational fluid dynamics models for proton exchange membrane fuel cell: a theoretical development. J Comput Multiph Flows 9(1):3–25

46. Martín-Alcántara A, Pino J, Iranzo A (2023) New insights into the temperature-water transport-performance relationship in PEM fuel cells. Int J Hydrog Energy 48(37):13987–13999

47. Wilberforce T, El-Hassan Z, Khatib FN, Al Makky A, Mooney J, Barouaji A, Olabi AG (2017) Development of Bi-polar plate design of PEM fuel cell using CFD techniques. Int J Hydrogen Energy 42(40):25663–25685

48. Yong Z, Shirong H, Xiaohui J, Mu X, Yuntao Y, Xi Y (2022) 3D multi-phase simulation of metal bipolar plate proton exchange membrane fuel cell stack with cooling flow field. Energy Convers Manage 273:116419

49. Ansys®, Fluent, release 2021 R1. Theory guide, chapter 25: Modeling Fuel Cells. ANSYS, Inc.

50. Iranzo A (2019) CFD applications in energy engineering research and simulation: an introduction to published reviews. Processes 7(12):883

51. Gada VH, Tandon MP, Elias J, Vikulov R, Lo S (2017) A large scale interface multi-fluid model for simulating multiphase flows. Appl Math Model 44:189–204. https://doi.org/10.1016/j.apm.2017.02.030

52. STAR-CCM+ documentation version 1.06, Siemens (2019)

53. Asanin S, Hinrichsen K, Lehnert W (2020) Water management in automotive polymer-electrolyte-membrane fuel cell stacks

54. Wang Y, Wang C-Y (2007) Two-phase transients of polymer electrolyte fuel cells. J Electrochem Soc 154(7):B636. https://doi.org/10.1149/1.2734076

55. Newman J, Thomas-Alyea KE (2004) Electrochemical systems, 3rd edn. John Wiley & Sons, Inc., p 519

56. Wesselingh JA, Krishna R (2000) Mass transfer in multicomponent mixtures. Delft University Press

57. Neyerlin KC, Gu W, Jorne J, Gasteiger HA (2006) Determination of catalyst unique parameters for the oxygen reduction reaction in a PEMFC. J Electrochem Soc 153(10)
58. Wang C-Y (2004) Fundamental models for fuel cell engineering. Chem Rev 104(10):4727–4765
59. Iranzo A, Muñoz M, Rosa F, Pino J (2010) Numerical model for the performance prediction of a PEM fuel cell. Model results and experimental validation. Int J Hydrog Energy 35(20):11533–11550. Iranzo A, Muñoz M, Pino J, Rosa F (2011) Update on numerical model for the performance prediction of a PEM Fuel Cell. Int J Hydrog Energy 36(15):9123–9127
60. Sui PC, Kumar S, Djilali N. Advanced computational tools for PEM fuel cell design. Part 2. Detailed experimental validation and parametric study. J Power Sources 180(1):423–432
61. Iranzo A, Boillat P, Rosa F (2014) Validation of a three dimensional PEM fuel cell CFD model using local liquid water distributions measured with neutron imaging. Int J Hydrog Energy 39(13):7089–7099
62. Tsushima S, Hirai S (2011) In situ diagnostics for water transport in proton exchange membrane fuel cells. Prog Energy Combust Sci 37(2):204–220
63. Iranzo A, Boillat P, Rosa F (2014) Validation of a three dimensional PEM fuel cell CFD model using local liquid water distributions measured with neutron imaging. Int J Hydrog Energy 39(13):7089–7099
64. Benkovic D, Fink C, Iranzo A. Qualitative and quantitative determination of liquid water distribution in a PEM fuel cell. Int J Hydrog Energy 52:1360–1370
65. AVL List GmbH, FIRE^{TM} User Manual 2023 R2 (2023)
66. Fink C (2009) Ph.D. Thesis, Graz University of Technology, Graz, Austria
67. Fink C, Fouquet N (2011) Electrochim Acta 56:10820
68. Karpenko-Jereb L, Innerwinkler P, Kelterer A-M, Sternig C, Fink C, Prenninger P, Tatschl R (2014) Int J Hydrogen Energy 39:7077
69. Karpenko-Jereb L, Sternig C, Fink C, Hacker V, Theiler A, Tatschl R (2015) J Power Sources 297:329
70. Fink C, Karpenko-Jereb L, Ashton S (2016) Fuel Cells 16:490
71. Karpenko-Jereb L, Sternig C, Fink C, Tatschl R (2016) Int J Hydrogen Energy 41:13644
72. Fink C, Kosir N, Tatschl R (2017) Proceeding of 6th European PEFC & Electrolyser forum, Lucerne, Switzerland, pp 24
73. Fink C, Gößling S, Karpenko-Jereb L, Urthaler P (2020) Fuel Cells 20:431
74. Benkovic D, Fink C, Iranzo A (2024) Int J Hydrogen Energy 52:1360
75. Fink C, Edjokola JM, Telenta M, Bodner M (2024) Fuel Cells, e202300237
76. Li S, Cao J, Wangard W, Becker U (2005) Modeling PEMFC with FLUENT: numerical performance and validations with experimental data. In: Proceedings of the ASME 2005 3rd international conference on fuel cell science, engineering and technology. 3rd international conference on fuel cell science, engineering and technology. Ypsilanti, Michigan, USA. May 23–25, pp 103–110. ASME. https://doi.org/10.1115/FUELCELL2005-74059
77. Iranzo A, Muñoz M, Rosa F, Pino J (2010) Numerical model for the performance prediction of a PEM fuel cell. Model results and experimental validation. Int J Hydrogen Energy 35(20):11533–11550. https://doi.org/10.1016/j.ijhydene.2010.04.129
78. Porstmann S, Wannemacher T, Drossel WG (2020) A comprehensive comparison of state-of-the-art manufacturing methods for fuel cell bipolar plates including anticipated future industry trends. J Manuf Process 60:366–383. Elsevier Ltd. https://doi.org/10.1016/j.jmapro.2020.10.041
79. Ansys fluent theory guide. Release 2024R1 (2024)
80. The fuel cell & electrolyzer module, COMSOL multiphysics v. 6.2. COMSOL AB, Stockholm, Sweden (2023)
81. Bockris JO'M, Reddy AKN, Gamboa-Aldeco M (2000) Modern electrochemistry, vol 2A, 2nd edn. Fundamentals of Electrodics, Kluwer Academic/Plenum Publishers, p 1052
82. Iranzo A, Gregorio JM, Boillat P, Rosa F (2020) Bipolar plate research using computational fluid dynamics and neutron radiography for proton exchange membrane fuel cells. Int J Hydrogen Energy 45(22):12432–12442. https://doi.org/10.1016/j.ijhydene.2020.02.183

83. Rafiq A, Rishikesh K, Vignarooban K, Kannan AM (2024) Optimized anode, cathode and coolant flow designs for enhanced performance of proton exchange membrane fuel cells. Int J Hydrogen Energy 50:1302–1313. https://doi.org/10.1016/j.ijhydene.2023.10.267

84. Ahmed R (2023) Optimized anode, cathode, coolant flow designs and optimization of operating conditions for enhanced performance of proton exchange membrane fuel cells. MS thesis. Arizona State University

85. Tolias IC, Giannissi SG, Venetsanos AG, Verbecke F, Molkov V (2019) Best practice guidelines in numerical simulations and CFD benchmarking for hydrogen safety applications. Int J Hydrogen Energy 9050–9062

86. Roache PJ (1998) Verification of codes and calculations. AIAA J 36(5):696–702

87. Iranzo A, Arredondo CH, Kannan AM, Rosa F (2020) Biomimetic flow fields for proton exchange membrane fuel cells: a review of design trends. Energy 190:116435

88. Tsotridis G, Pilenga A, De Marco G, Malkow T (2015) EU harmonised test protocols for PEMFC MEA testing in single cell configuration for automotive applications. JRC Science for Policy report. EUR 27632 EN. https://doi.org/10.2790/54653